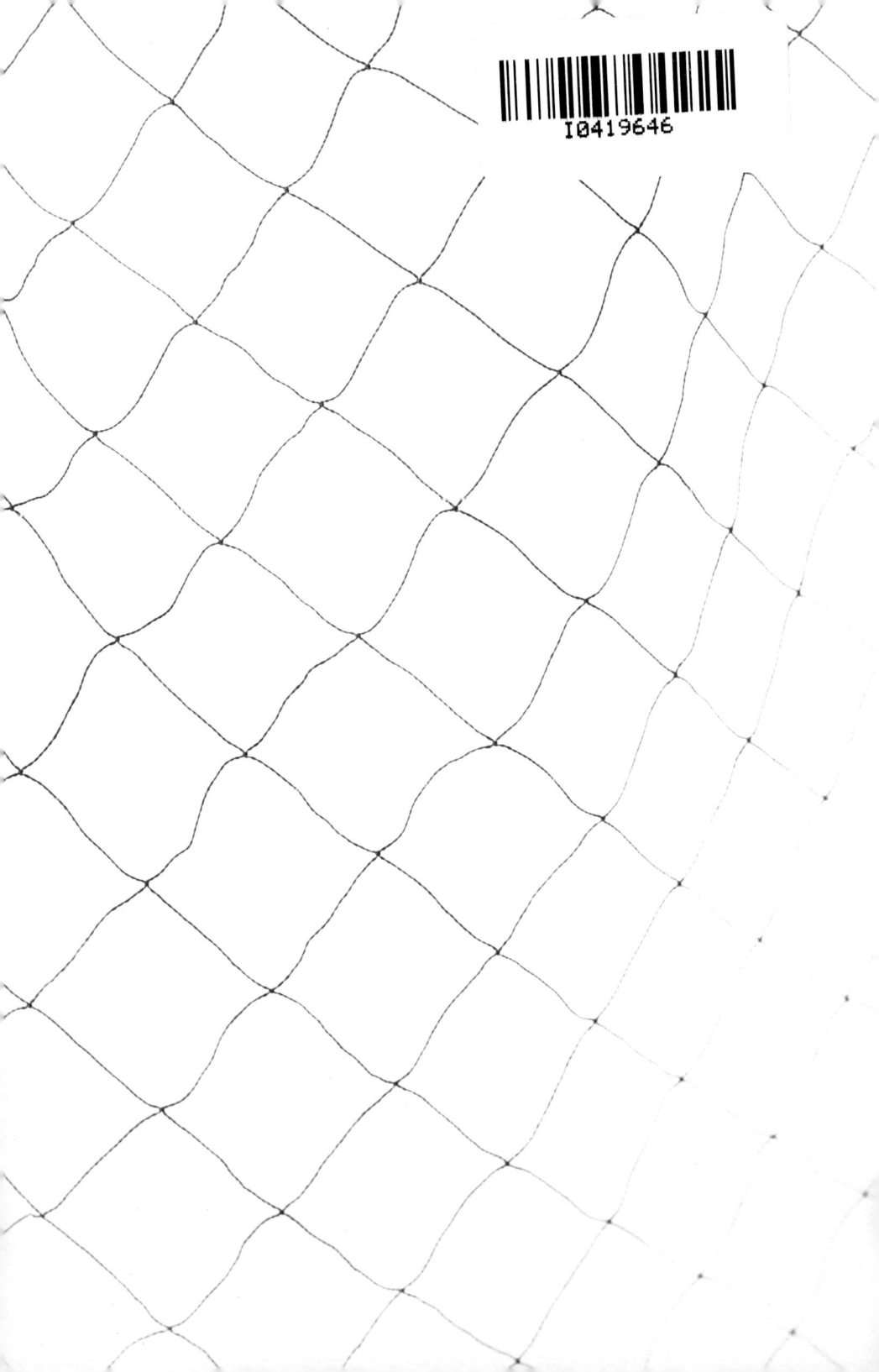

I0419646

Todd R. Forsgren

ORNITHOLOGICAL PHOTOGRAPHS

Daylight

Cofounders: Taj Forer and Michael Itkoff
Designer: Ursula Damm
Illustrations: Julian Montague
Editing: Alice Lovejoy
Copy editor: Elizabeth Bell

© 2015 Daylight Community Arts Foundation

Photographs © Todd R. Forsgren, 2006–2014

Introduction © Brian W. Forsgren, 2015
"A Brief History of Bird-banding and Mist-netting" © James Lowen, 2015
"Not Strictly for the Birds" © John A. Tyson, 2015
"Snaring the Viewer" © Susan Wegner, 2015

ISBN 978-1-942084-06-8

Printed in China

Daylight Books
E-mail: info@daylightbooks.org
Web: www.daylightbooks.org

TABLE OF CONTENTS

INTRODUCTION
Brian W. Forsgren

I.

My son Todd's connection to birds stems from a serendipitous twist of academic-administrative fate. In 1973, when I was in graduate school at the Ohio State University's Department of Zoology, my advisor needed a field guide for a bird-watching course. Somehow I was assigned this task, although I didn't know a Wood Duck from a domestic chicken. Given the circumstances, I made up my mind to be the best field guide I could be. The experience led me to become an ornithology field leader for the next three years, with my wife, Suzanne (Todd's remarkable mother), an eager partner in the process. For both of us, this sparked a lifelong love of birds and their incredible natural histories and survival strategies.

As life went on, Suzanne and I dragged our children into our passion for birding. Family vacations and outings revolved around it. New locations led to new species and new habitats, and deepened our appreciation of the natural world. A wonder for avian life was imprinted in Todd's sensibilities as his tender eyes saw a world around him that was invisible to most. He drank it in, becoming extremely focused and even a bit obsessed—going so far as to compete in bird-watching competitions and, in high school, attend bird-watching camp! In these photographs, I am proud to see the way Todd's extraordinary capacity for observation, reflection, and attention to detail have taken these early interests in an astonishing and complicated new direction.

II.

In my work as a veterinarian, I see literally hundreds of animals per week, many of them in desperate physical condition. But I never overlook their owner's desperation. In these clinical scenarios, the power of this union is evident.

I believe that this human-animal bond is an intrinsic part of human nature, that animals connect us, *Homo sapiens,* to an inherent aspect of ourselves—an aspect that is threatened by our ever increasing distance from the natural world. Our earliest artistic impulses seem to testify to our need to be part of nature and to the human-animal bond. Picture the Chauvet-Pont-d'Arc cave paintings in France, where some images date back as far as 32,000 years. Imagine ancient humans making the critical decision to adorn this cave with drawings of bison, horses, and bears. At least thirteen species are depicted, all of them likely important features in the daily life of that ancient world. It's impossible to know exactly why early humans made these drawings. Did they consider the animals to be gods? Were they used for ritual purposes, or for storytelling? We will never know for sure. But we do know that some of these creatures were food sources—and others, predators that stalked early human beings.

Todd's photographs update this tradition by reflecting on the complexity of this relationship between humans and other animals in our globalized world. His birds are caught in a state of cognitive dissonance. Their facial expressions are at times terrified, at times disgusted, and at times angry at the absurdity of their plight. In seeing a bird trapped in this way, do we empathize? Are the birds caught in a system similar to the one that entraps so many of us? The natural world is facing challenges such as industrialization and globalization, which are stretching the limits of survival strategies developed over millions of years of evolution. Yet it seems that Todd's birds recognize the perils ahead. They are the canaries in the coal mine of our often misguided treatment of the wondrous natural world, in which the complicated side effects of industrialized life, such as climate change and ecosystem destruction, put all creatures in peril. We need to make substantial changes for their sake, and our own.

III.

As a veterinarian, I'm forced to make some very difficult decisions. Similarly, ecologists make hard choices based on abstract values related to conservation, such as diversity and the survival of endangered species. Todd spent his formative years reflecting on things like this. I remember how, as a young boy, he processed the ethical ramifications of a program to maximize the survival rate of Kirtland's Warbler (a critically endangered species). A crucial part of this program is "controlling" the population of Brown-headed Cowbirds, a species that lays eggs in the nests of other birds. Cowbirds have had a devastating impact on the Kirtland's Warbler's survival, to the point that, at one time, fewer than 500 warblers remained. The warbler's recovery was only possible through the elimination of many times that number of cowbirds. It's an imperfect solution for an imperfect world. Although Todd's photographs celebrate the remarkable beauty and diversity of bird species, they also represent this sort of challenging ethical encounter.

I feel lucky to have participated in bird-banding operations like those Todd has photographed. I have had the privilege of holding a terrified creature in my hands, ever so gently. Feeling its tiny heart pounding is comparable to the moment when I first held Todd as an infant. For me, the most rewarding and certainly most emotional part of both parenting and bird-banding is the release. The bird in my hand looks at me with eyes filled with anger, even fury. As I open my fingers and the bird darts out, I feel a special responsibility for this unique creature—an empathy that translates into a sense of responsibility for all creatures. In the process of banding, I worry about a bird's individual fragility, yet this practice is justified by risk-benefit analysis: I know that this temporary inconvenience yields information that might help the entire species survive. Todd's images challenge us to think more critically and abstractly about how to be better stewards of the natural world.

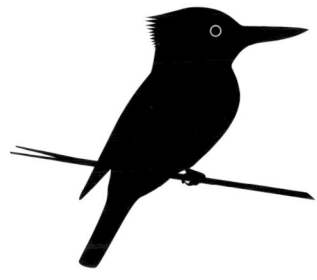

NON-PASSERINES
doves to woodpeckers

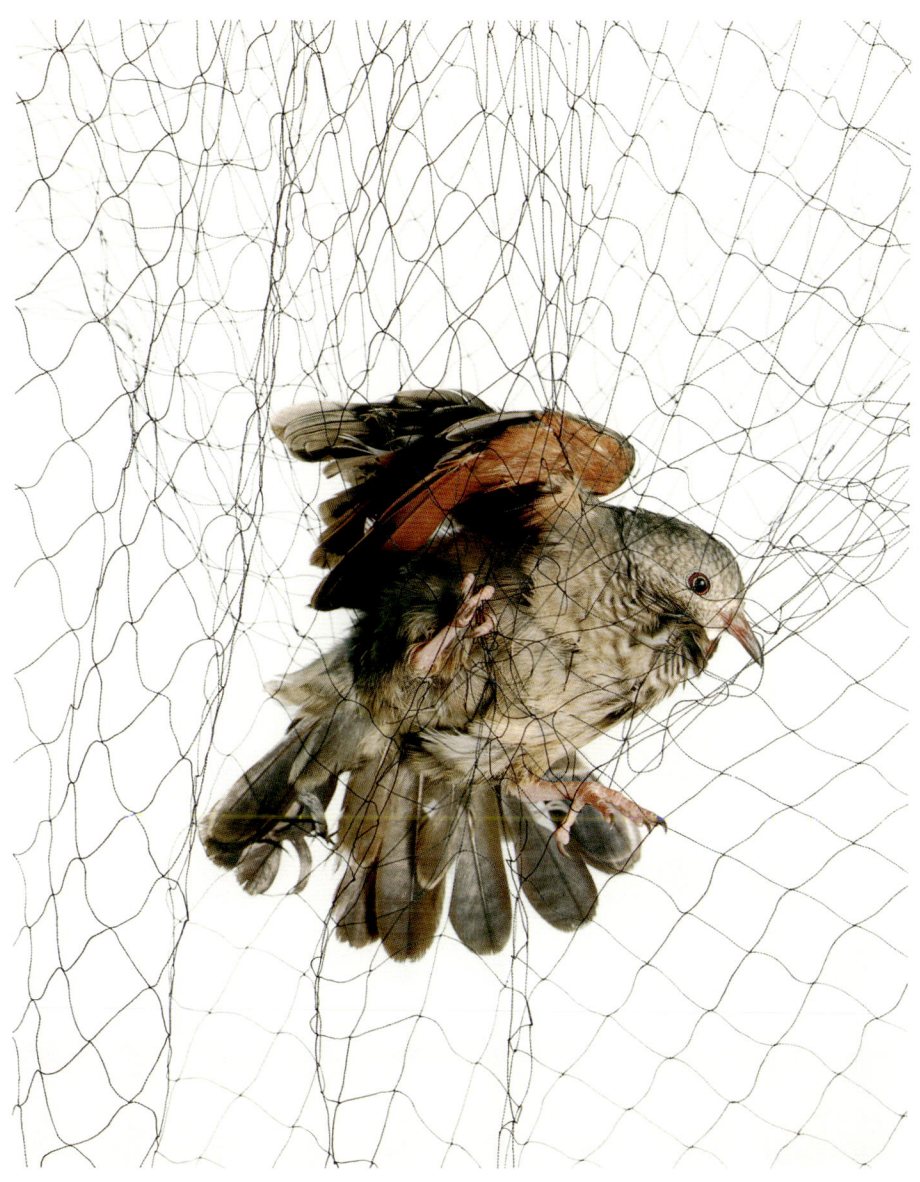

Common Ground-dove
Columbina passerina

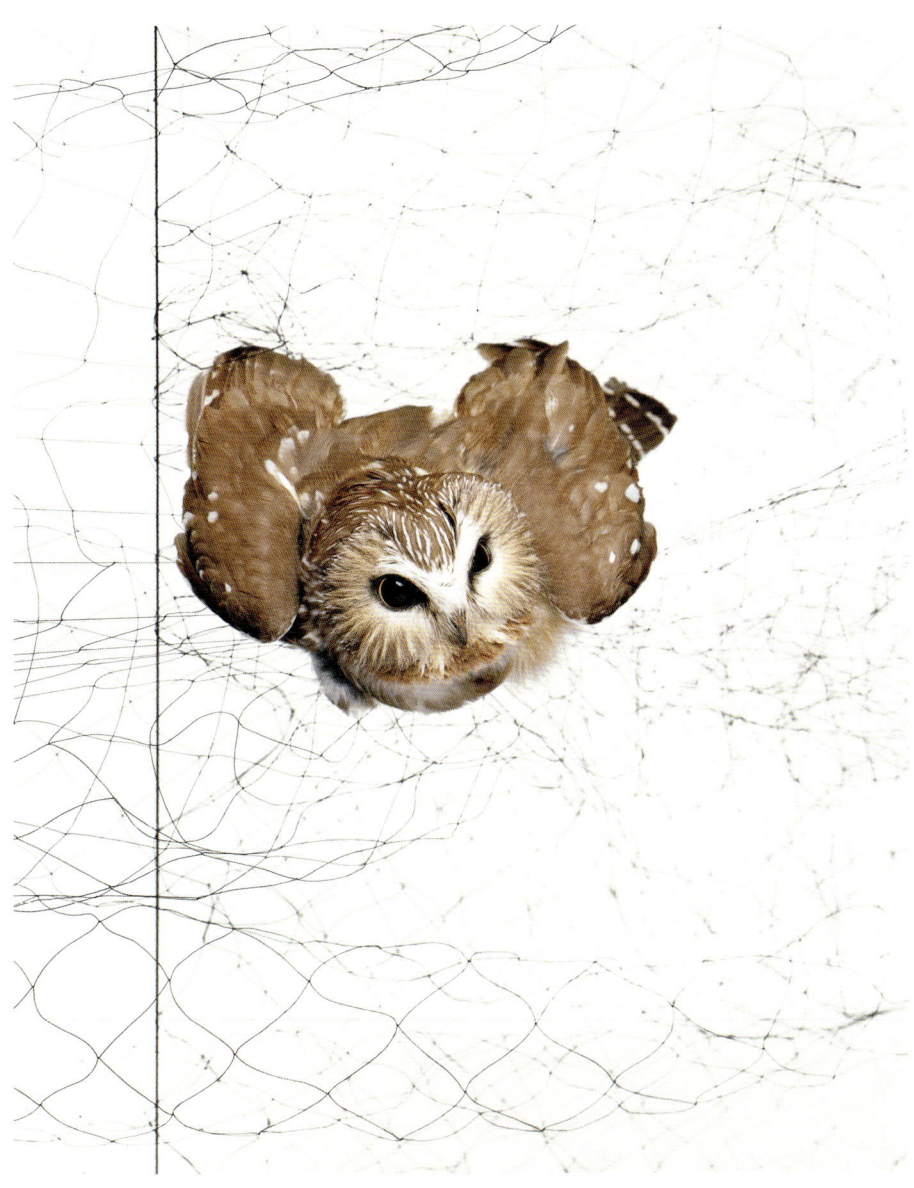

Northern Saw-whet Owl
Aegolius acadicus

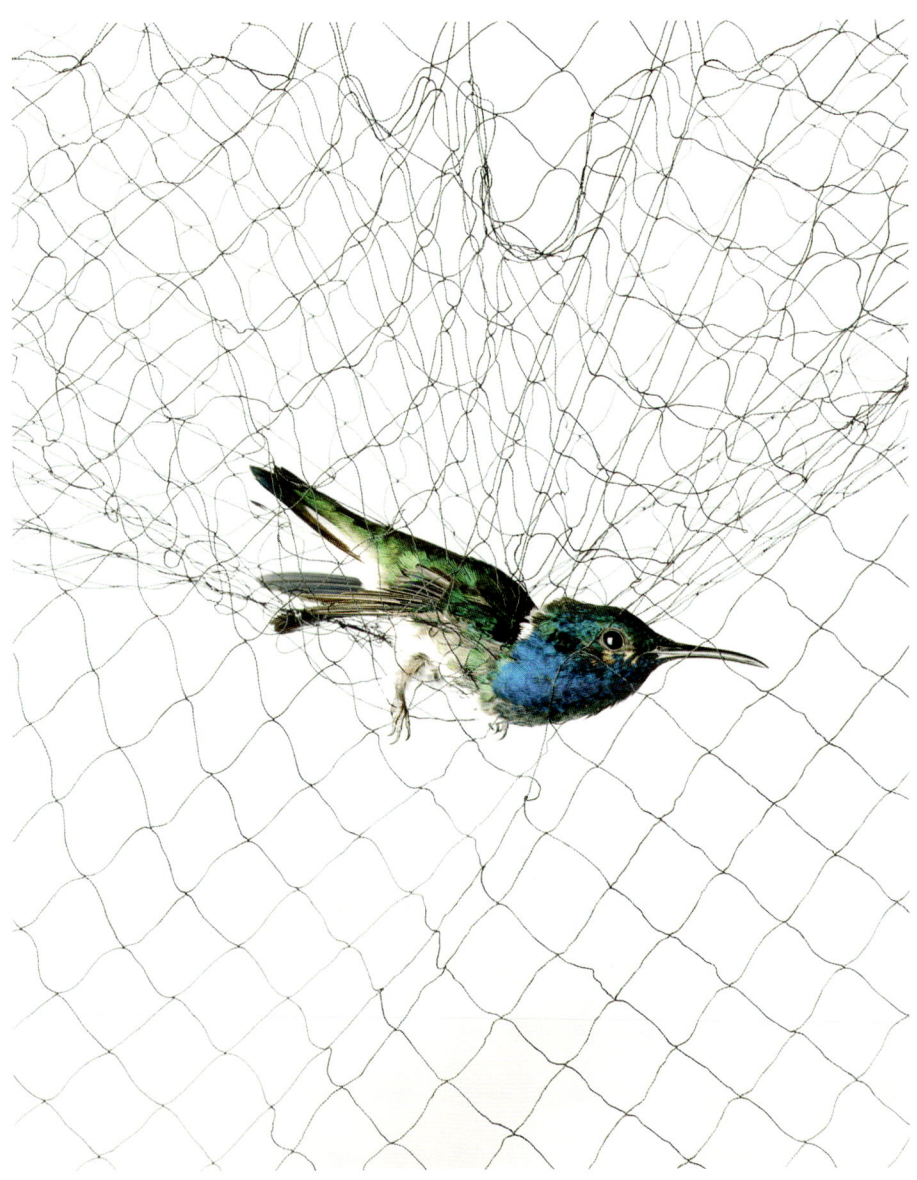

White-necked Jacobin
Florisuga mellivora

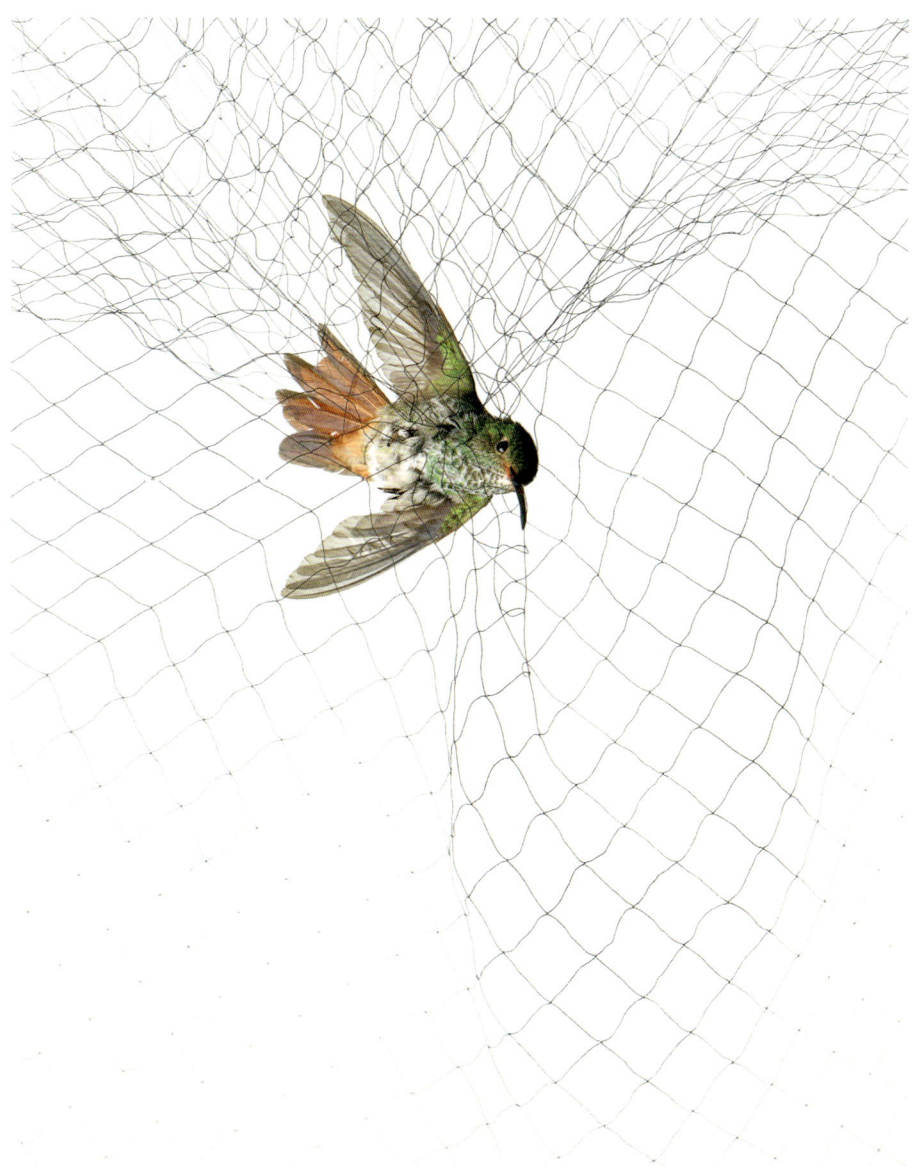

Rufous-tailed Hummingbird
Amazilia tzacatl

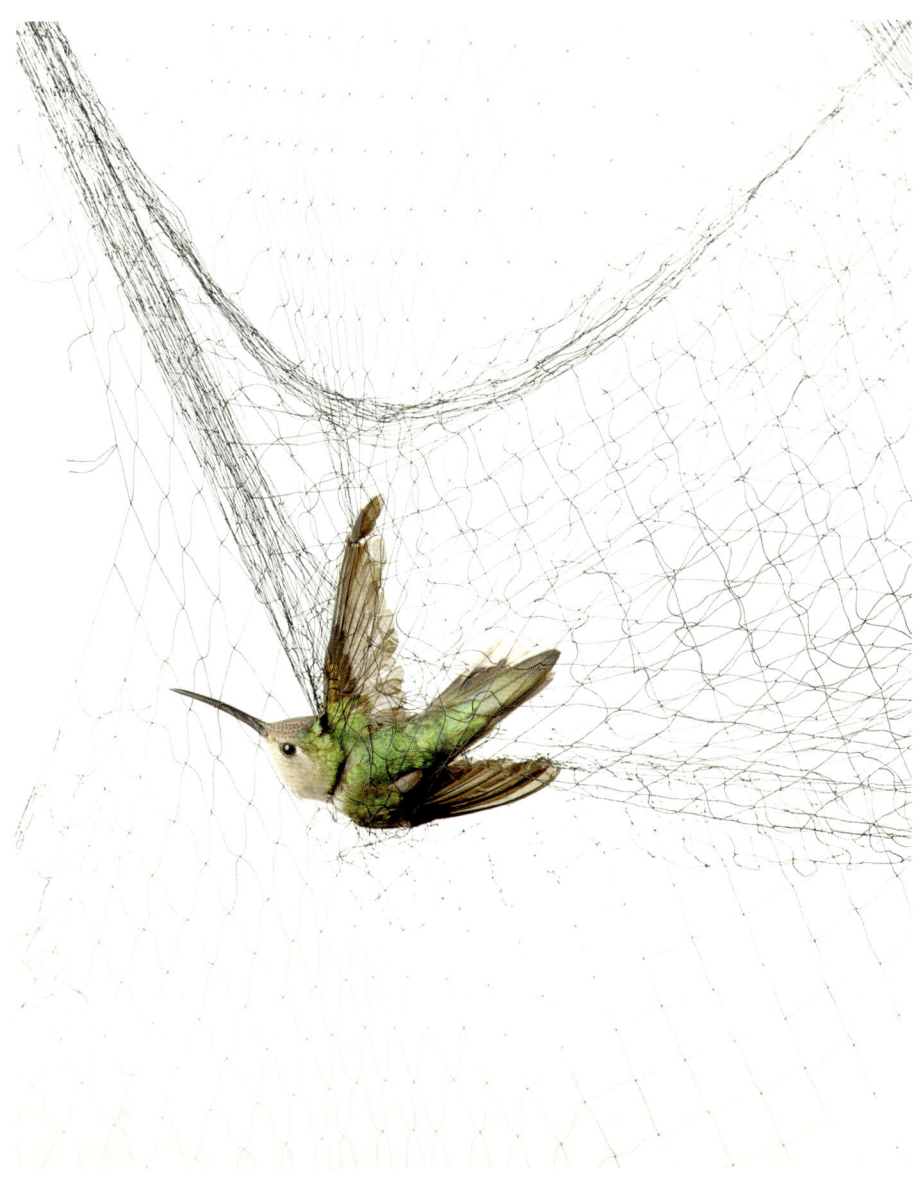

Violet-crowned Hummingbird
Amazilia violiceps

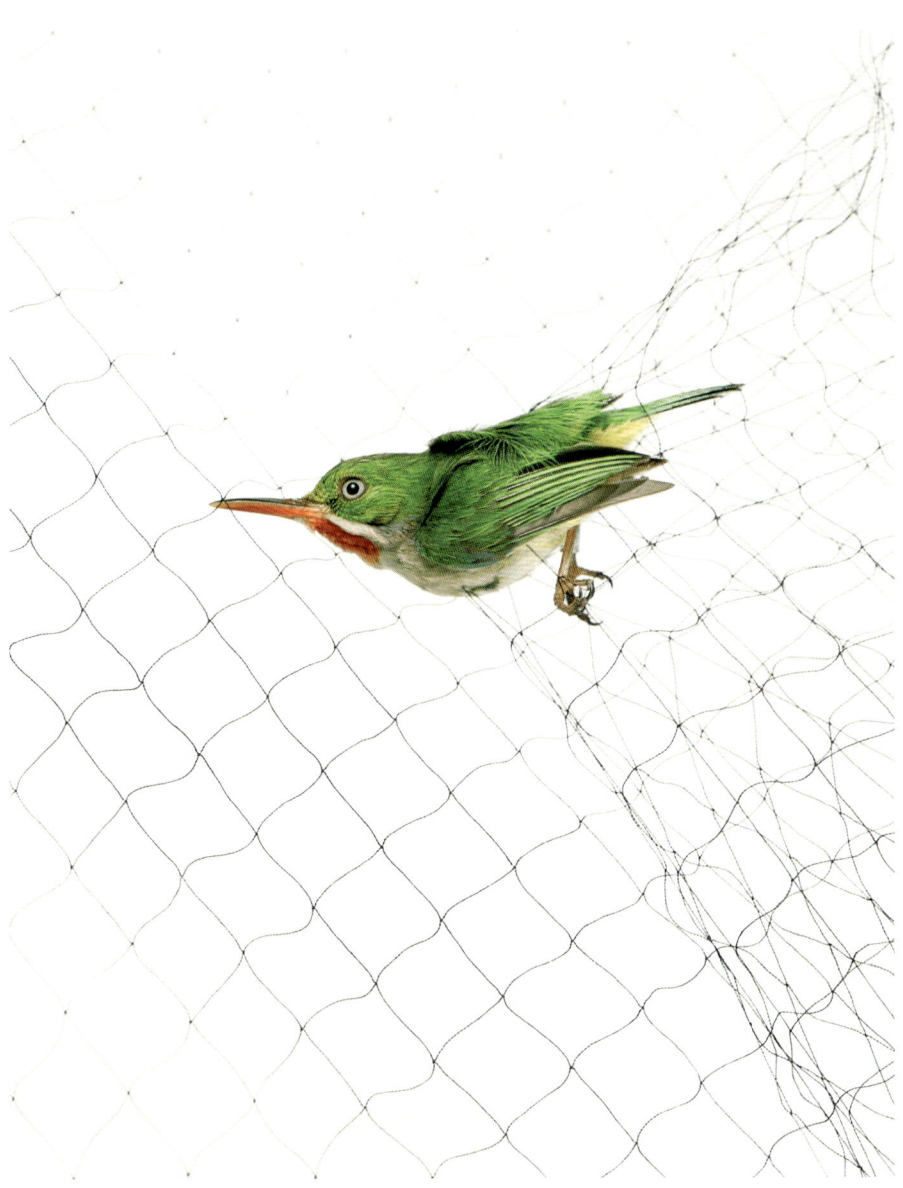

Puerto Rican Tody
Todus mexicanus

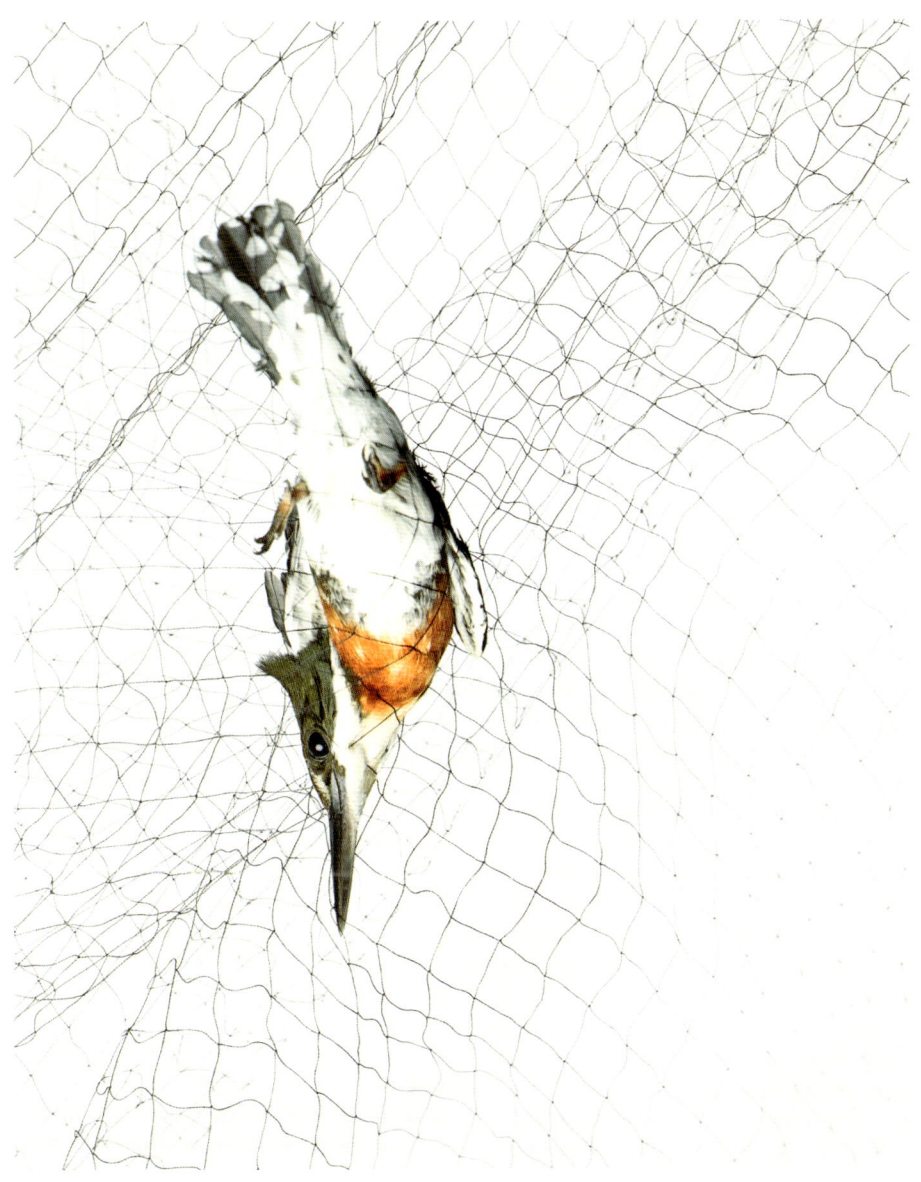

Green Kingfisher
Chloroceryle americana

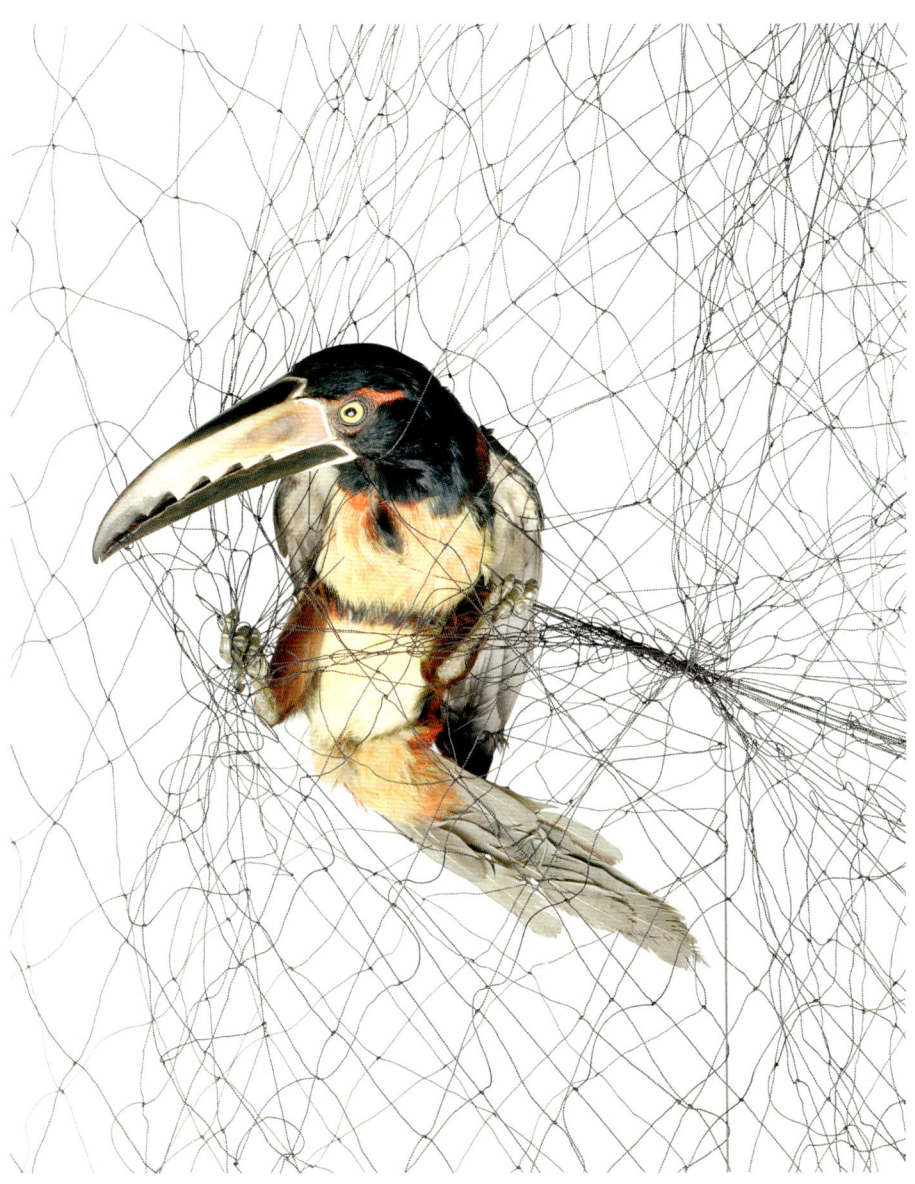

Collared Araçari
Pteroglossus torquatus

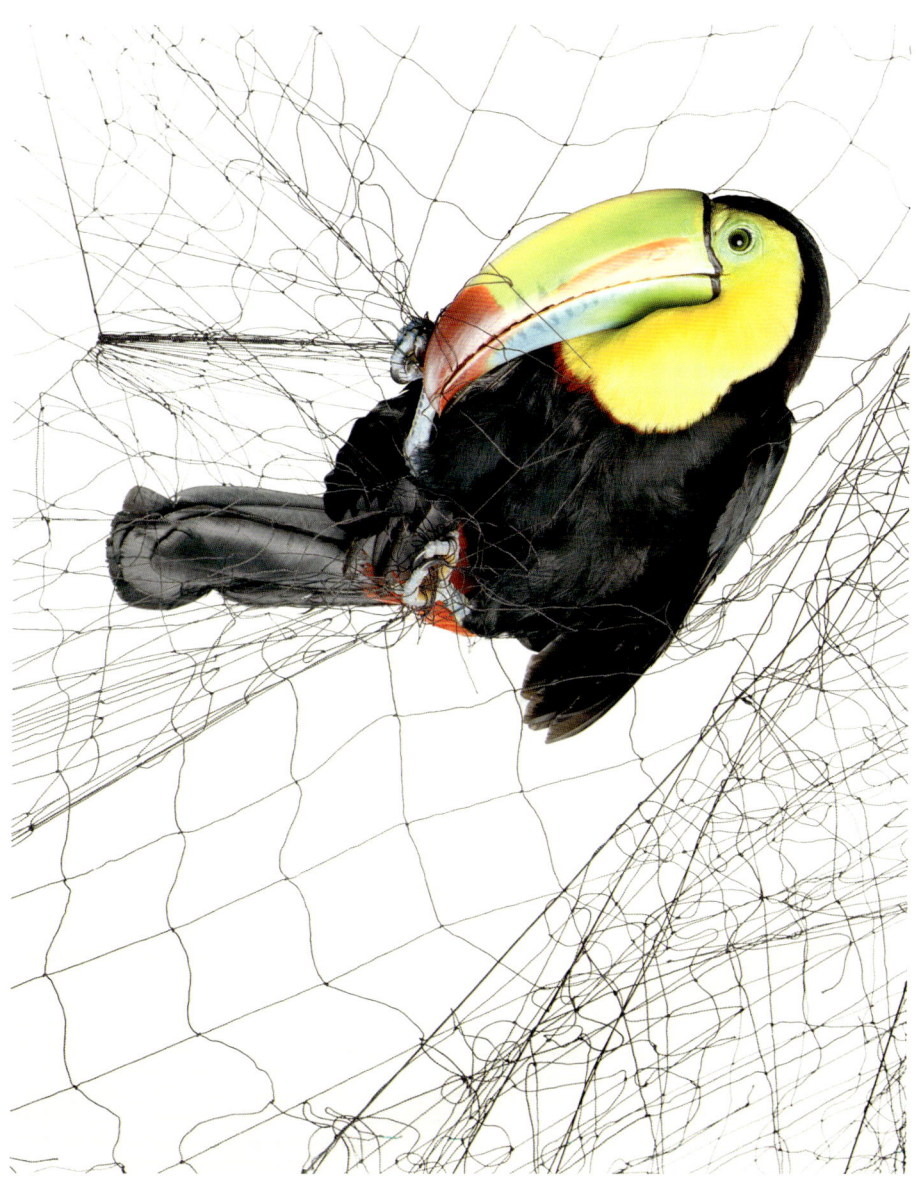

Keel-billed Toucan
Ramphastos sulfuratus

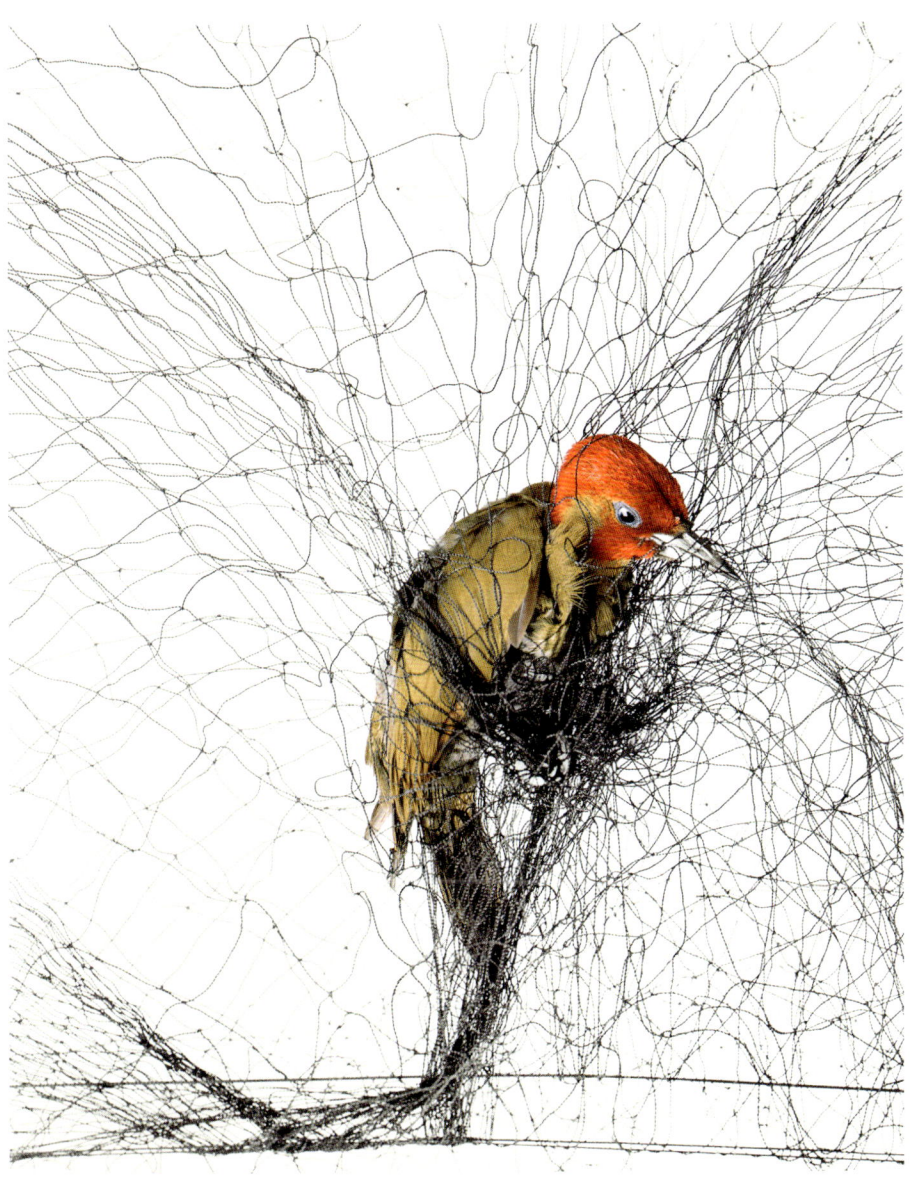

Rufous-winged Woodpecker
Piculus simplex

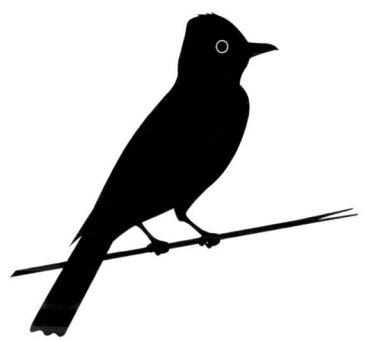

PASSERINES
Sub-oscines
antbirds to becards

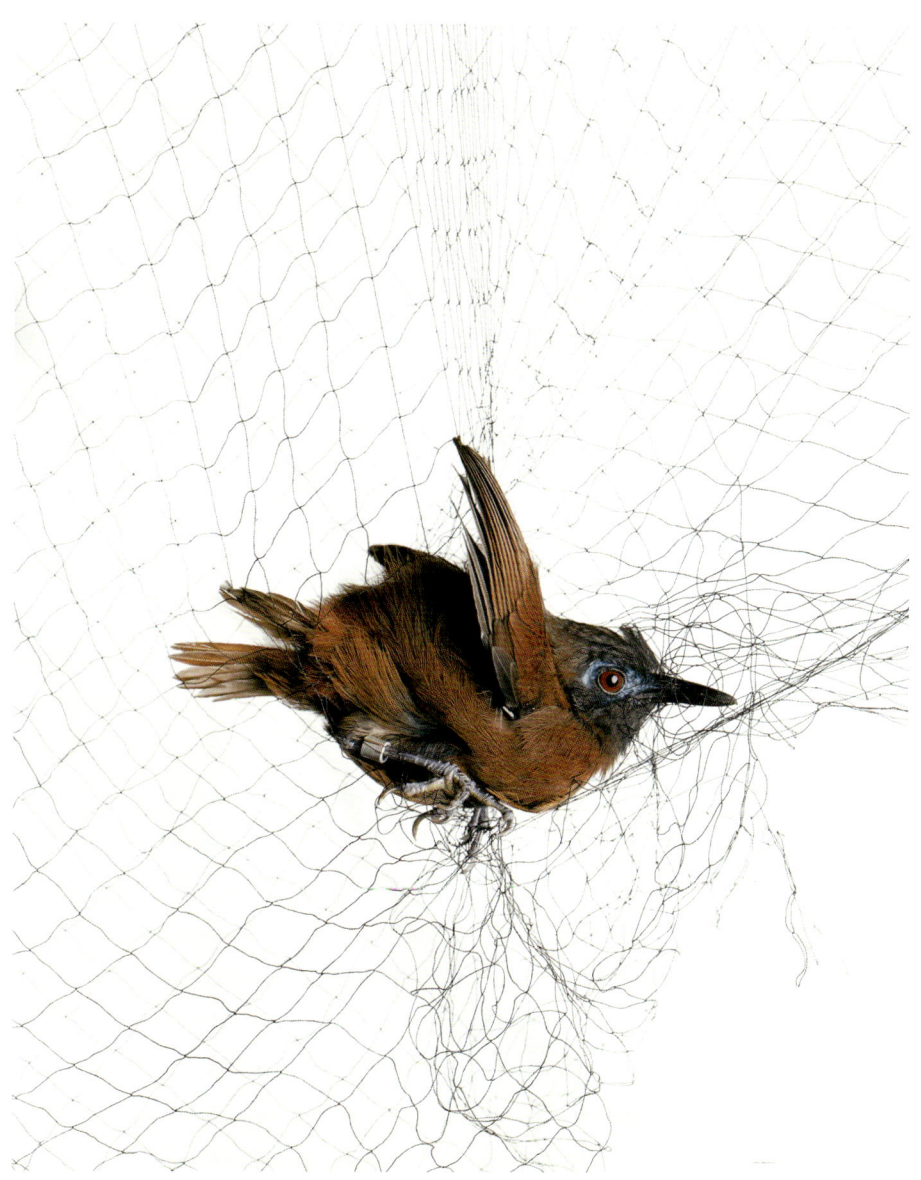

Zeledon's Antbird
Myrmeciza zeledonia

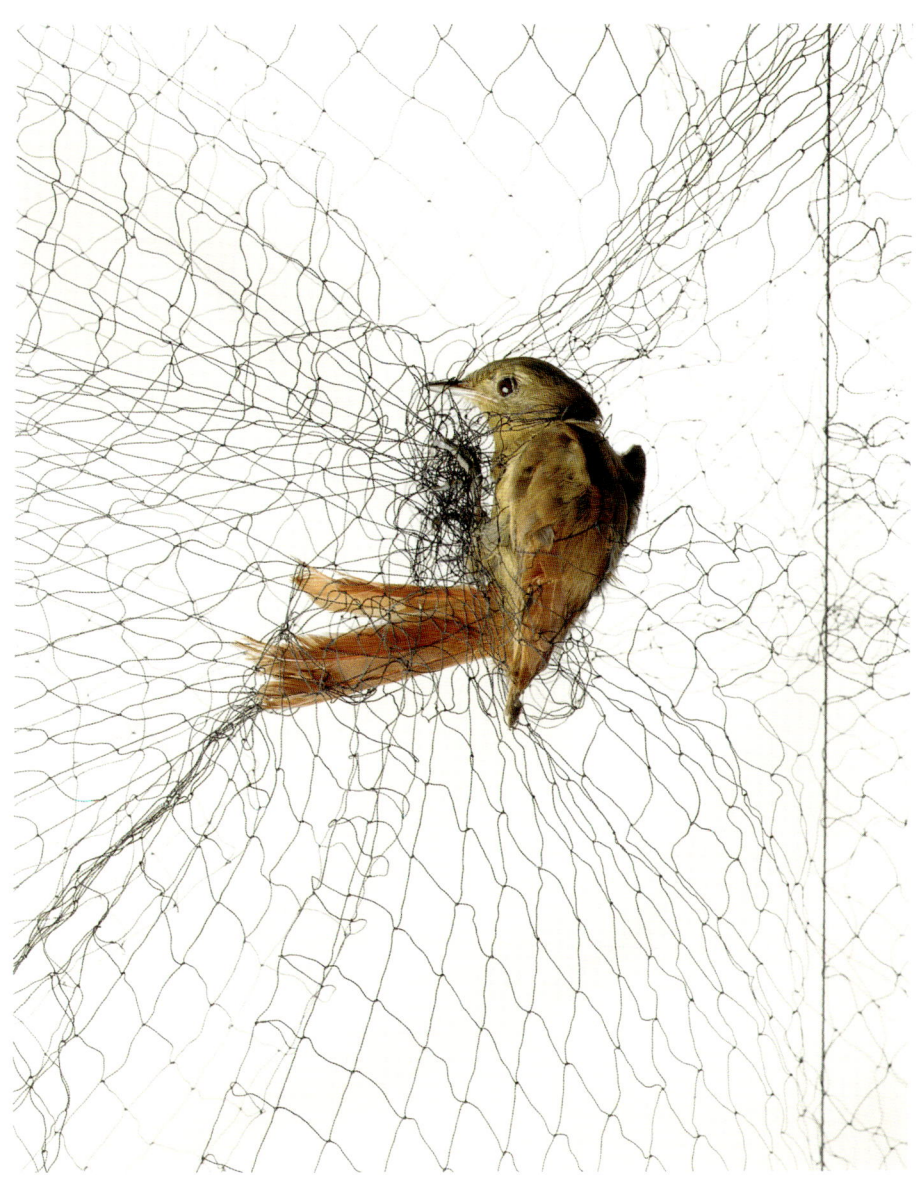

Olivaceous Woodcreeper
Sittasomus griseicapillus

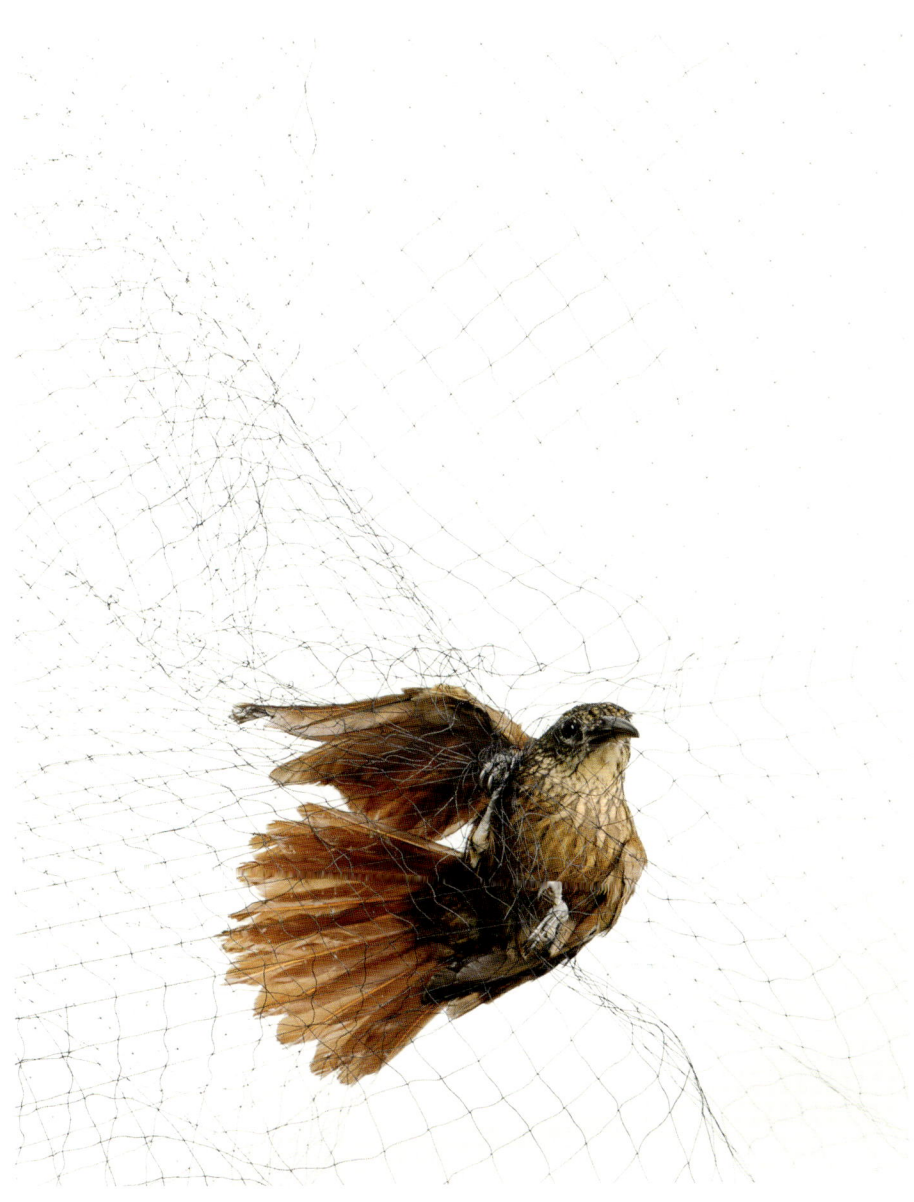

Cocoa Woodcreeper
Xiphorhynchus susurrans

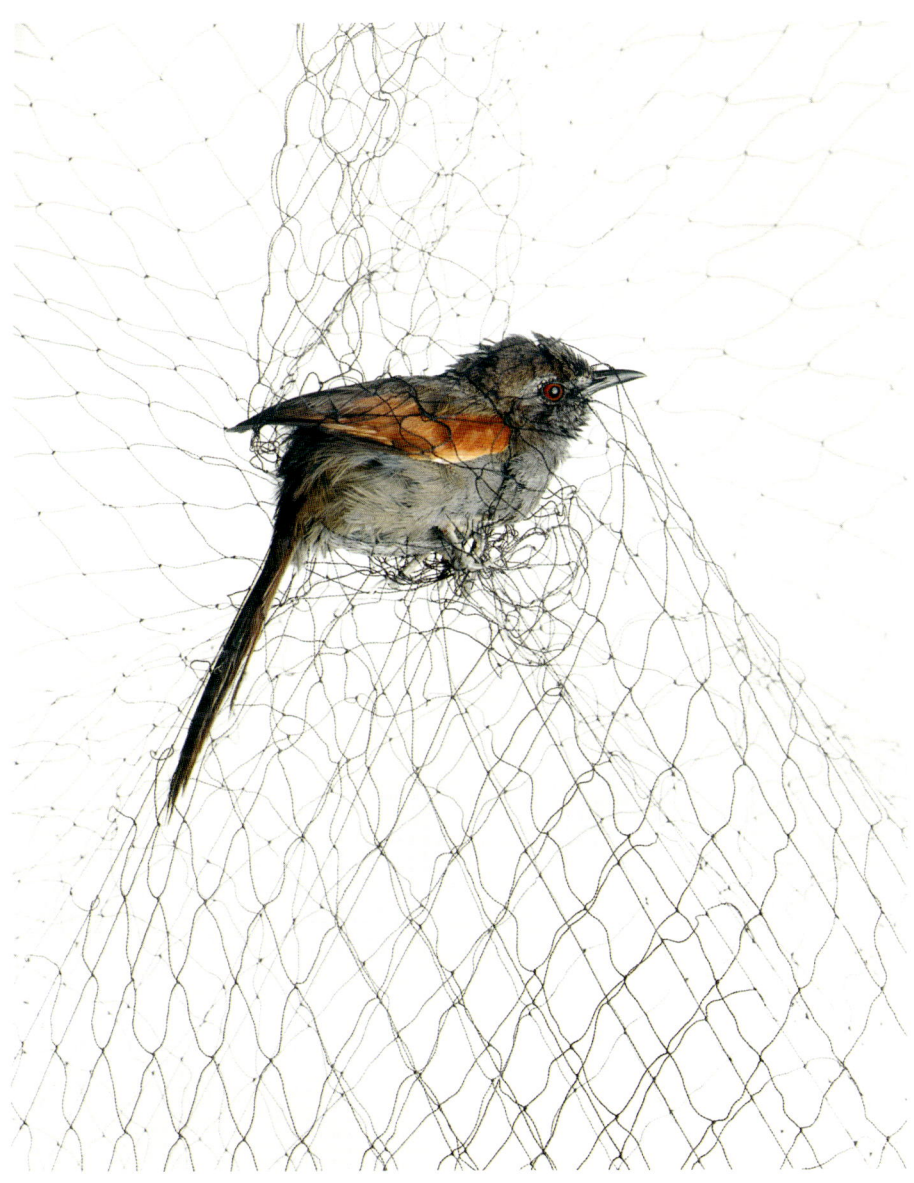

Gray-bellied Spinetail
Synallaxis cinerascens

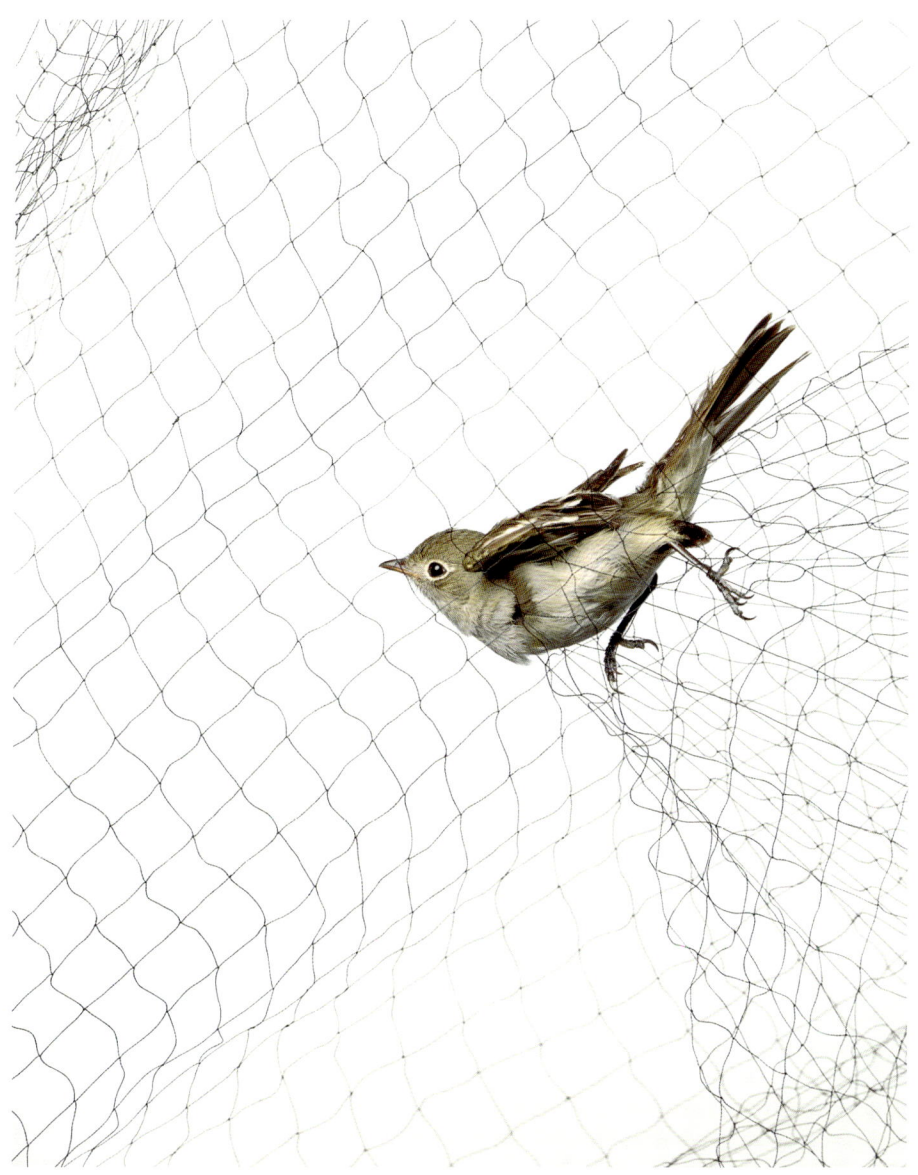

Small-billed Elaenia
Elaenia parvirostris

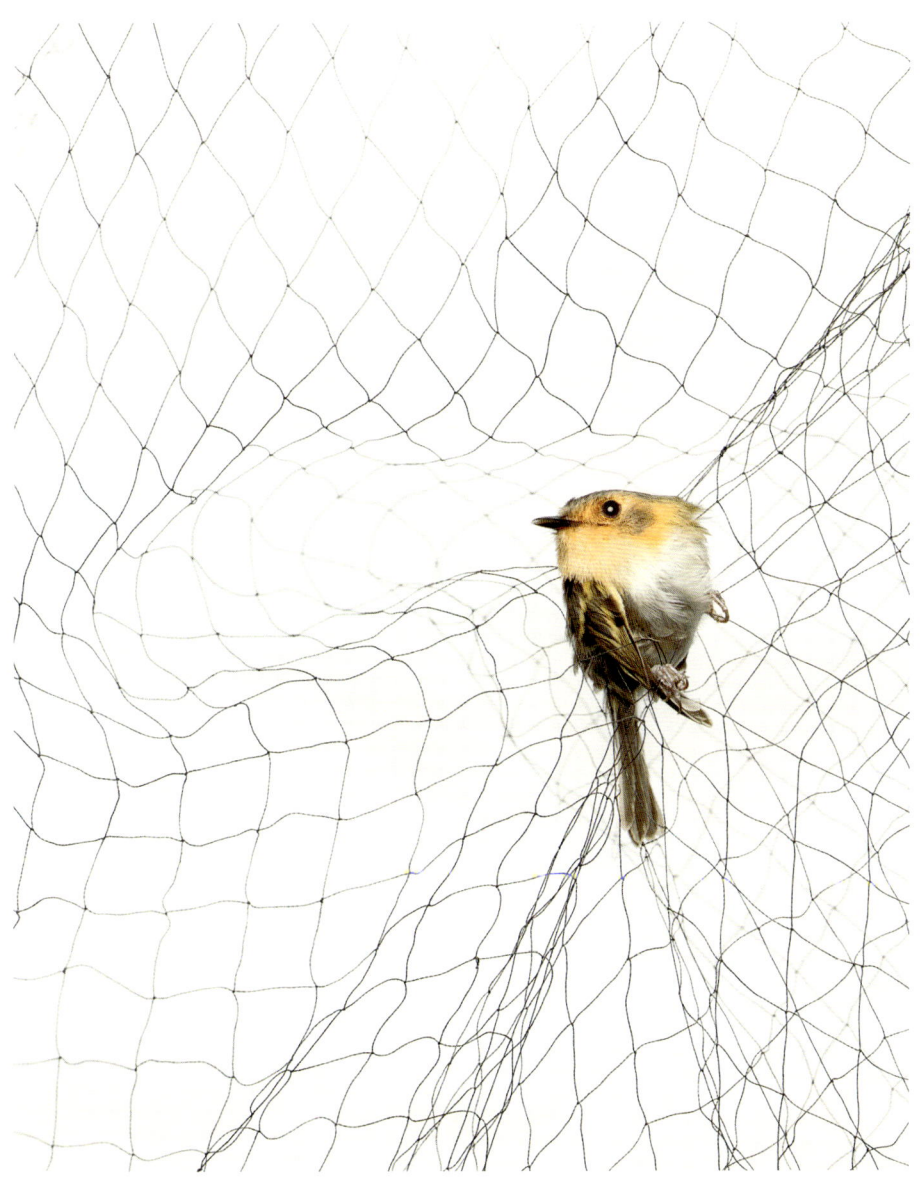

Ochre-faced Tody-flycatcher
Poecilotriccus plumbeiceps

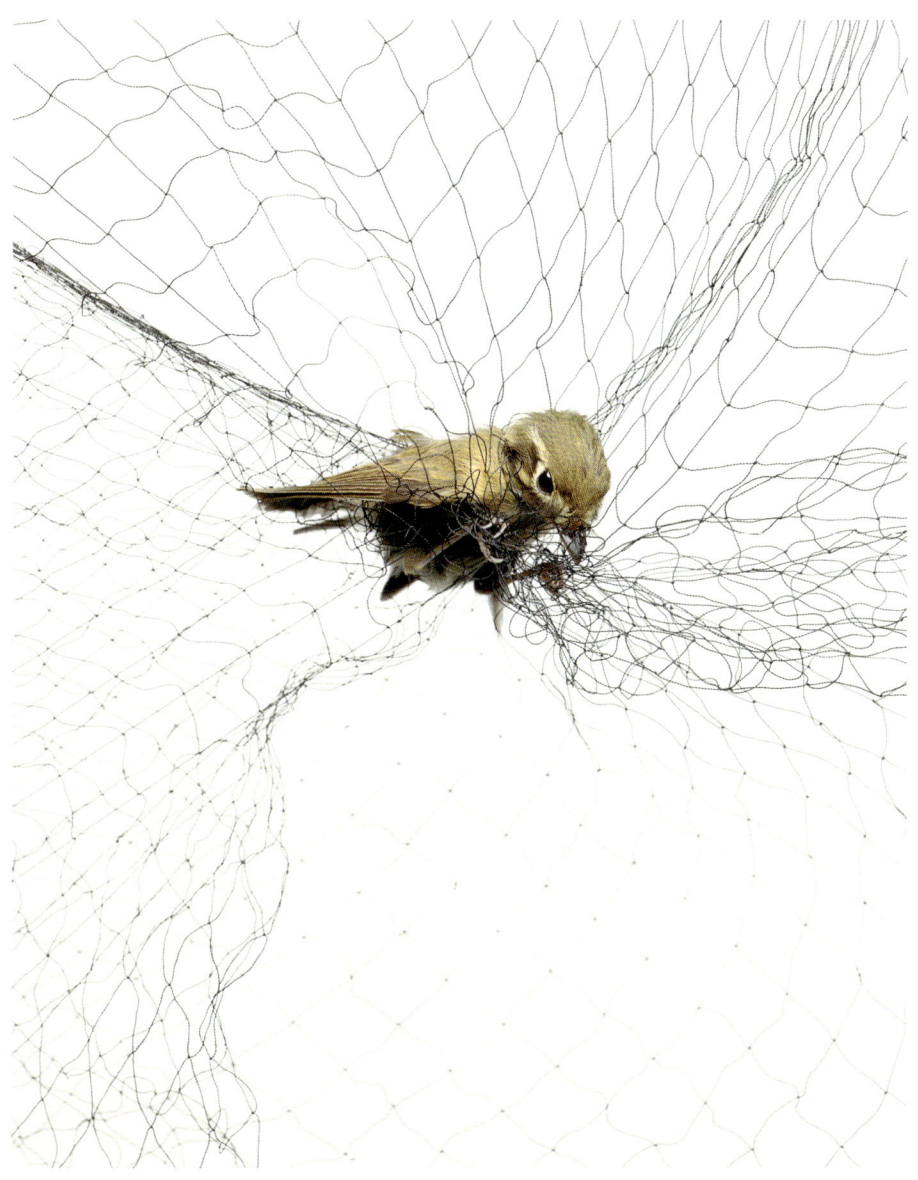

White-throated Spadebill
Platyrinchus mystaceus

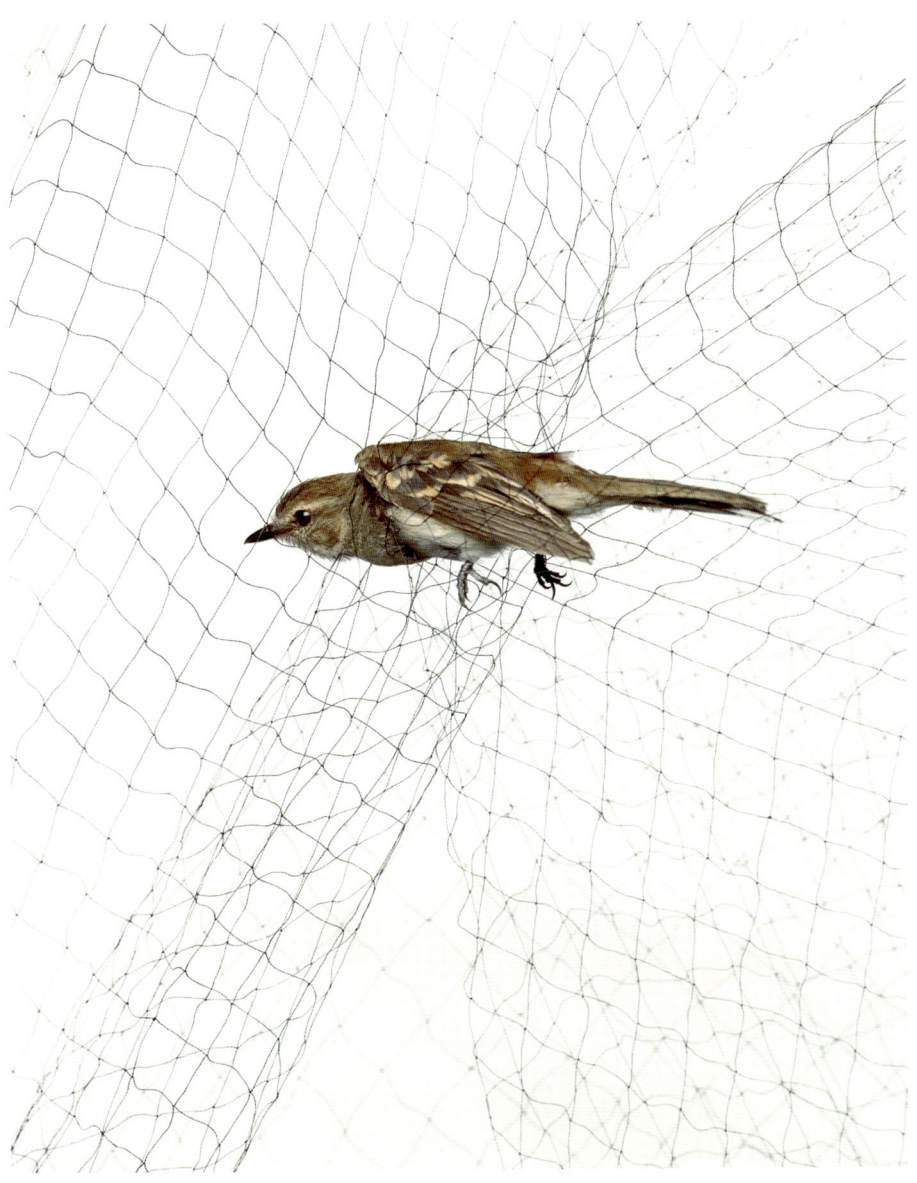

Fuscous Flycatcher
Cnemotriccus fuscatus

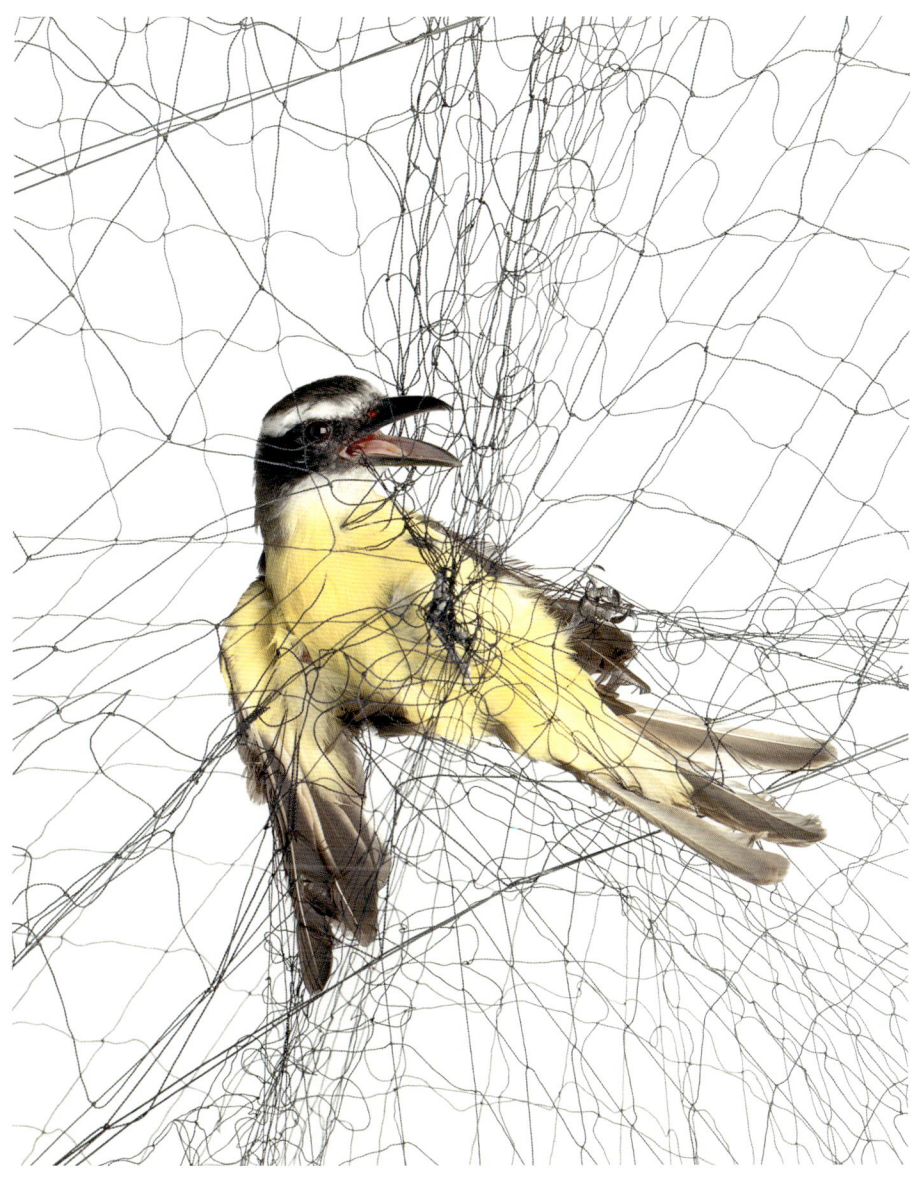

Boat-billed Flycatcher
Megarynchus pitangua

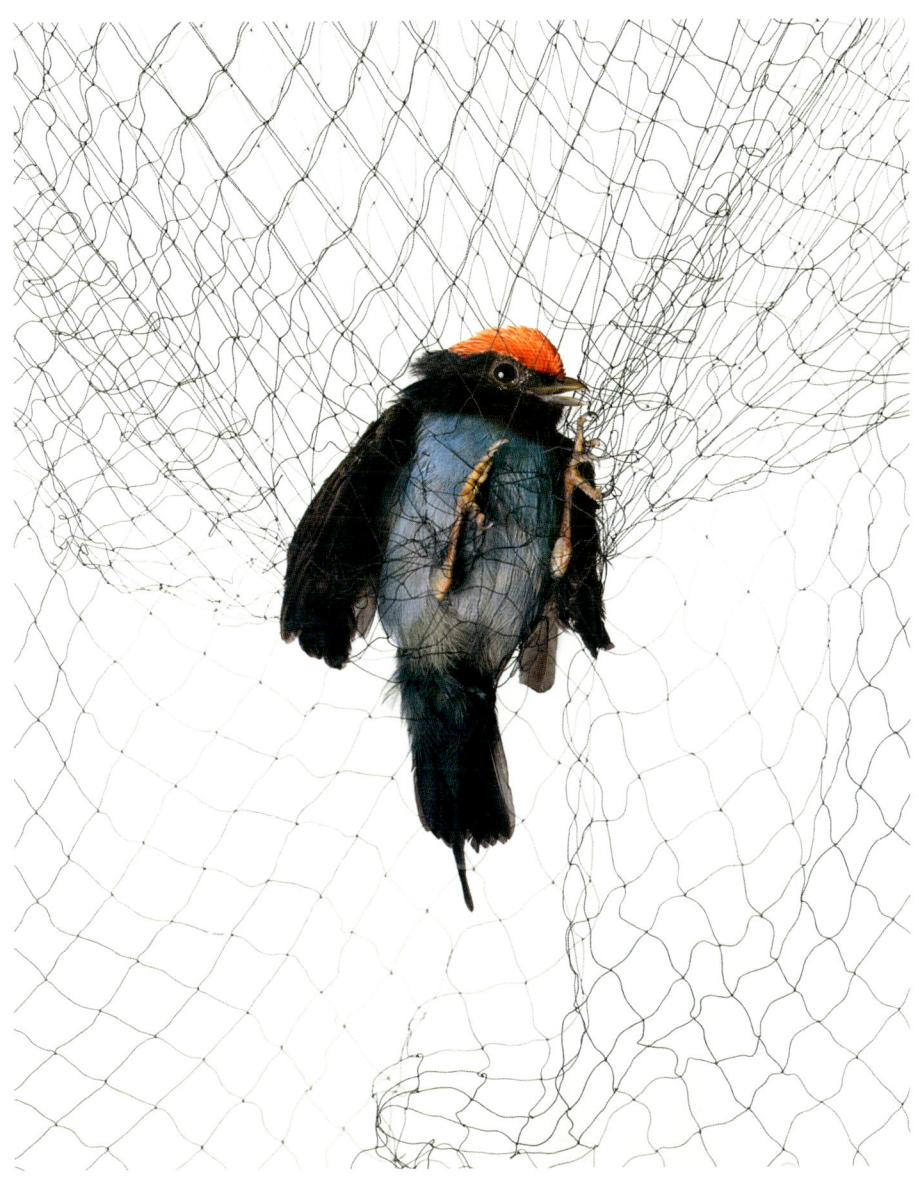

Swallow-tailed Manakin
Chiroxiphia caudata

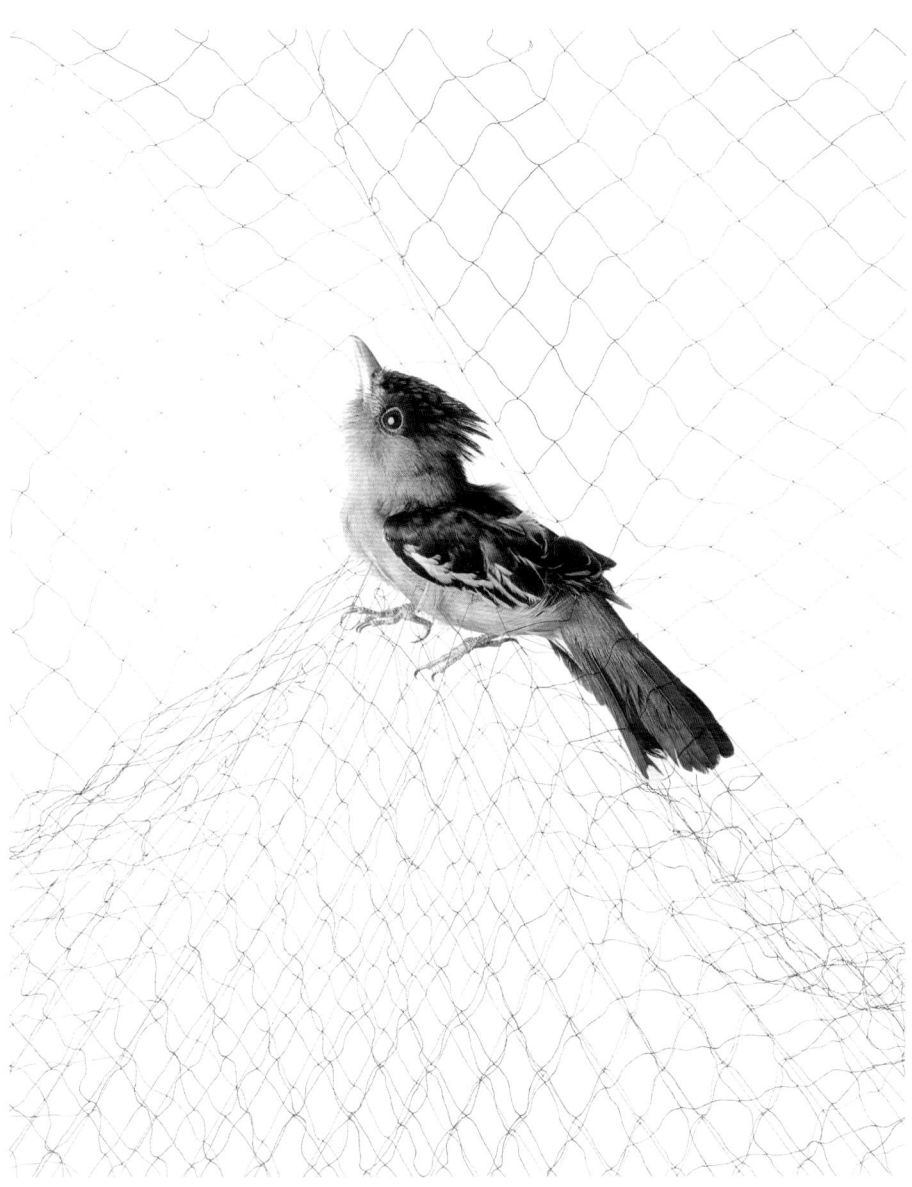

White-winged Becard
Pachyramphus polychopterus

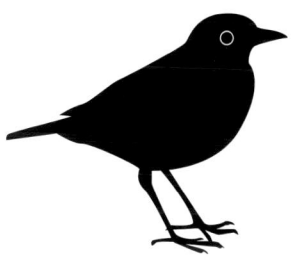

PASSERINES
Oscines
vireos to silky-flycatchers

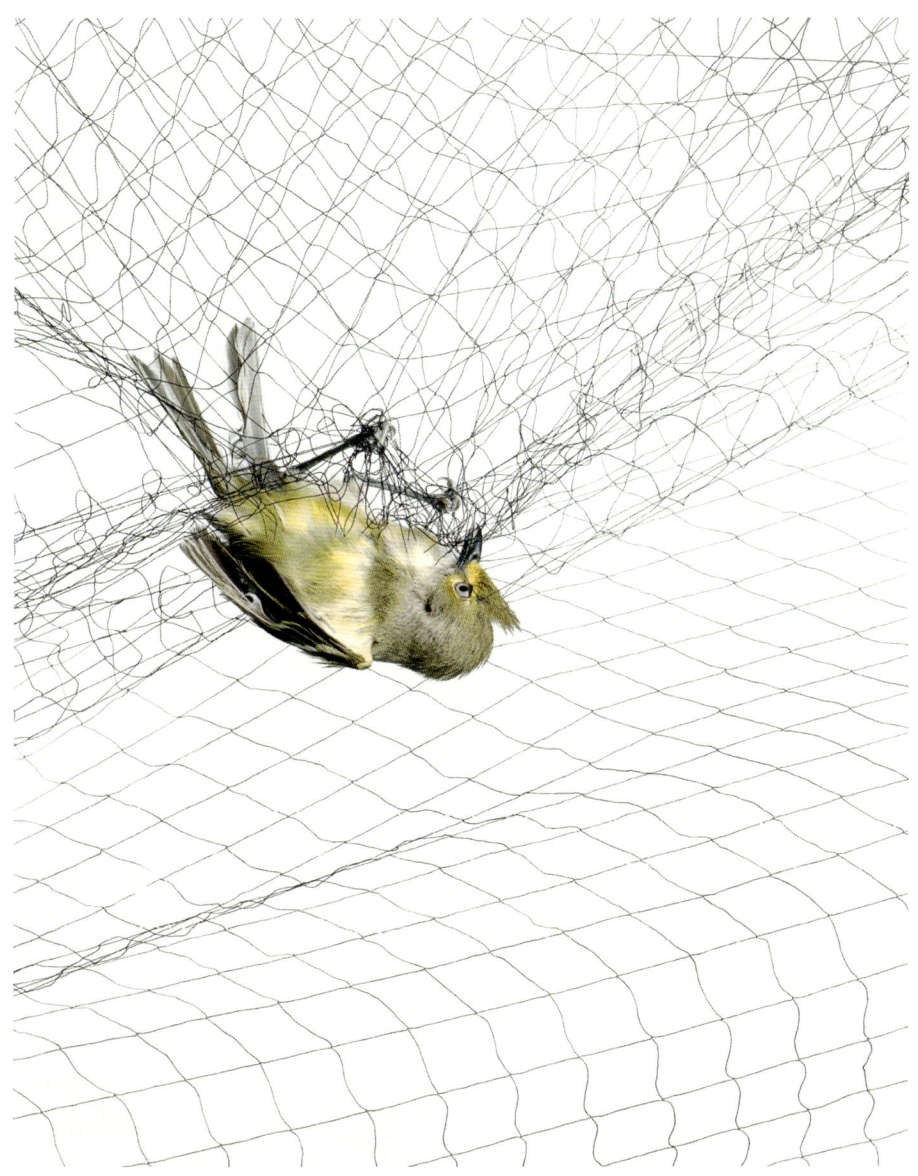

White-eyed Vireo
Vireo griseus

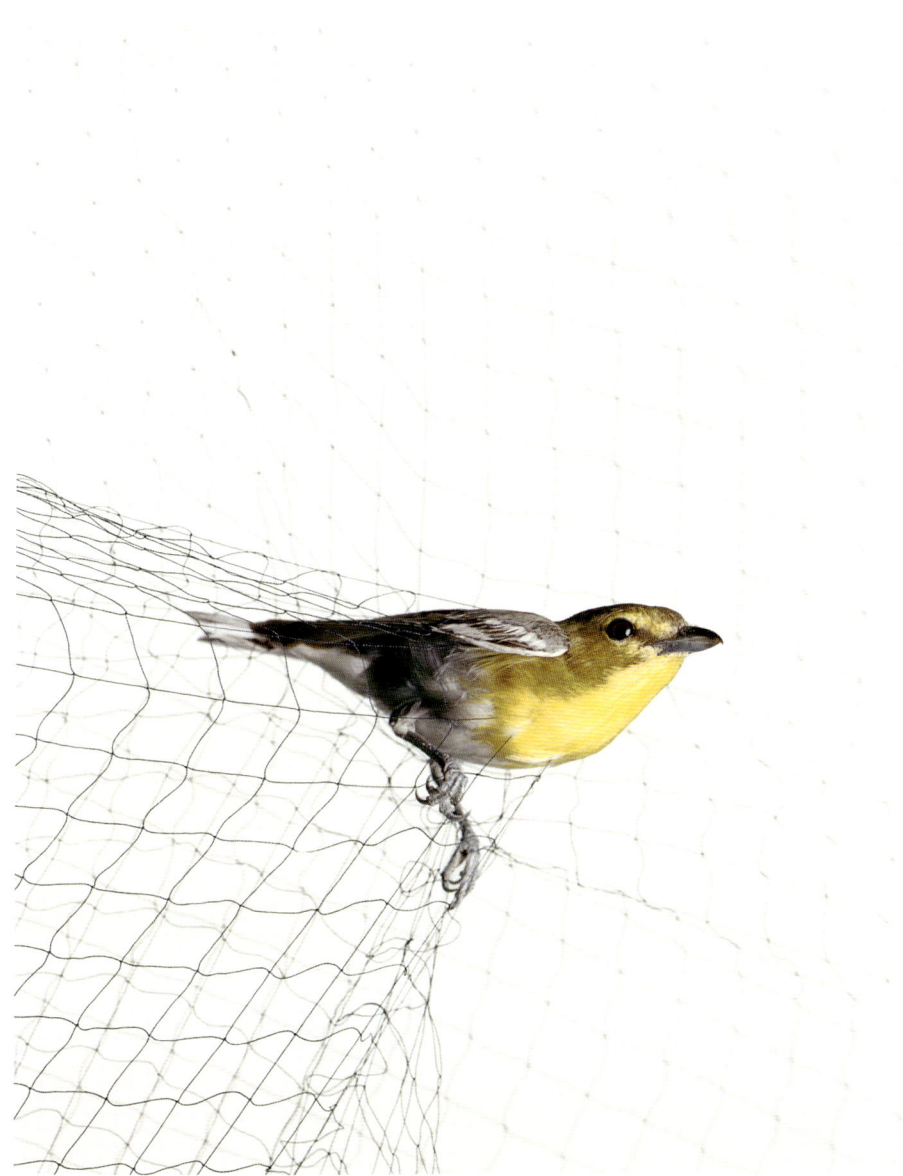

Yellow-throated Vireo
Vireo flavifrons

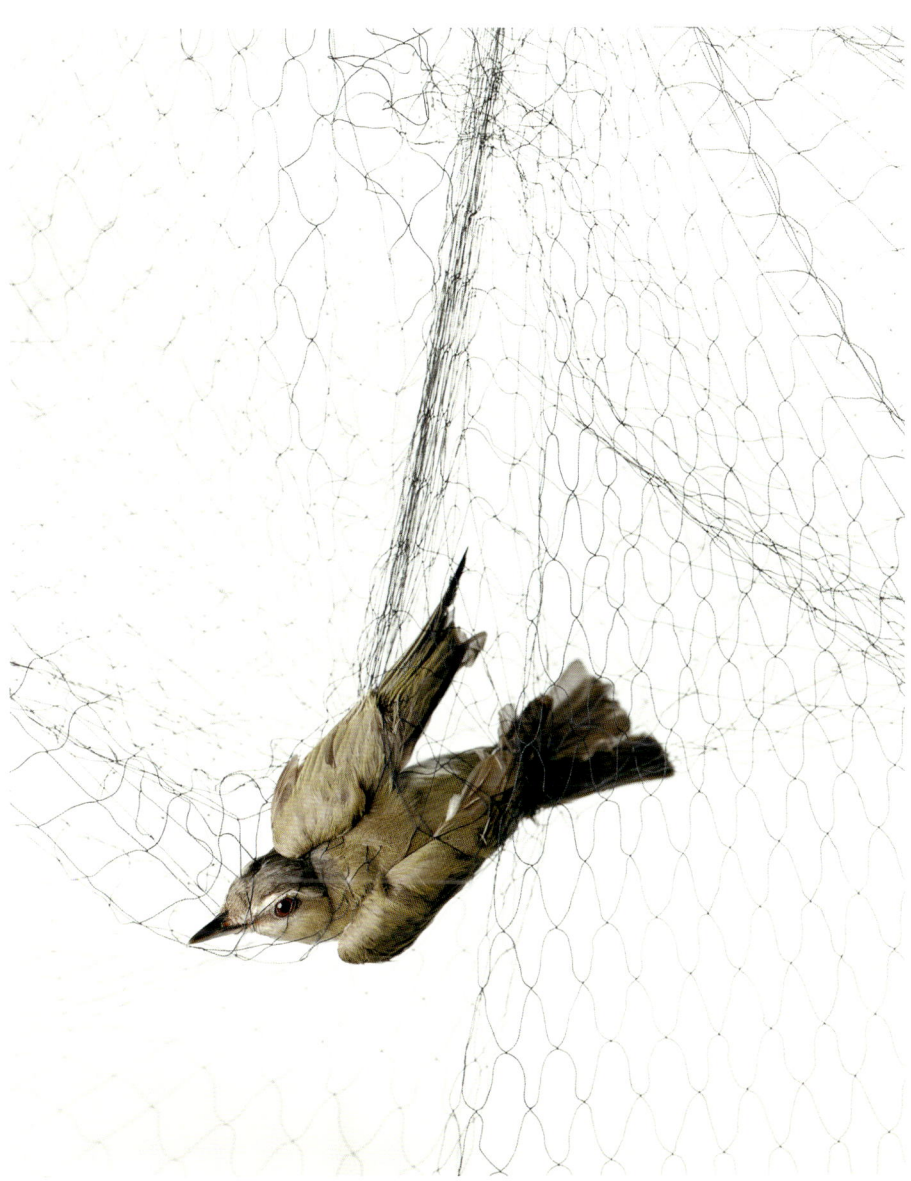

Red-eyed Vireo
Vireo olivaceus

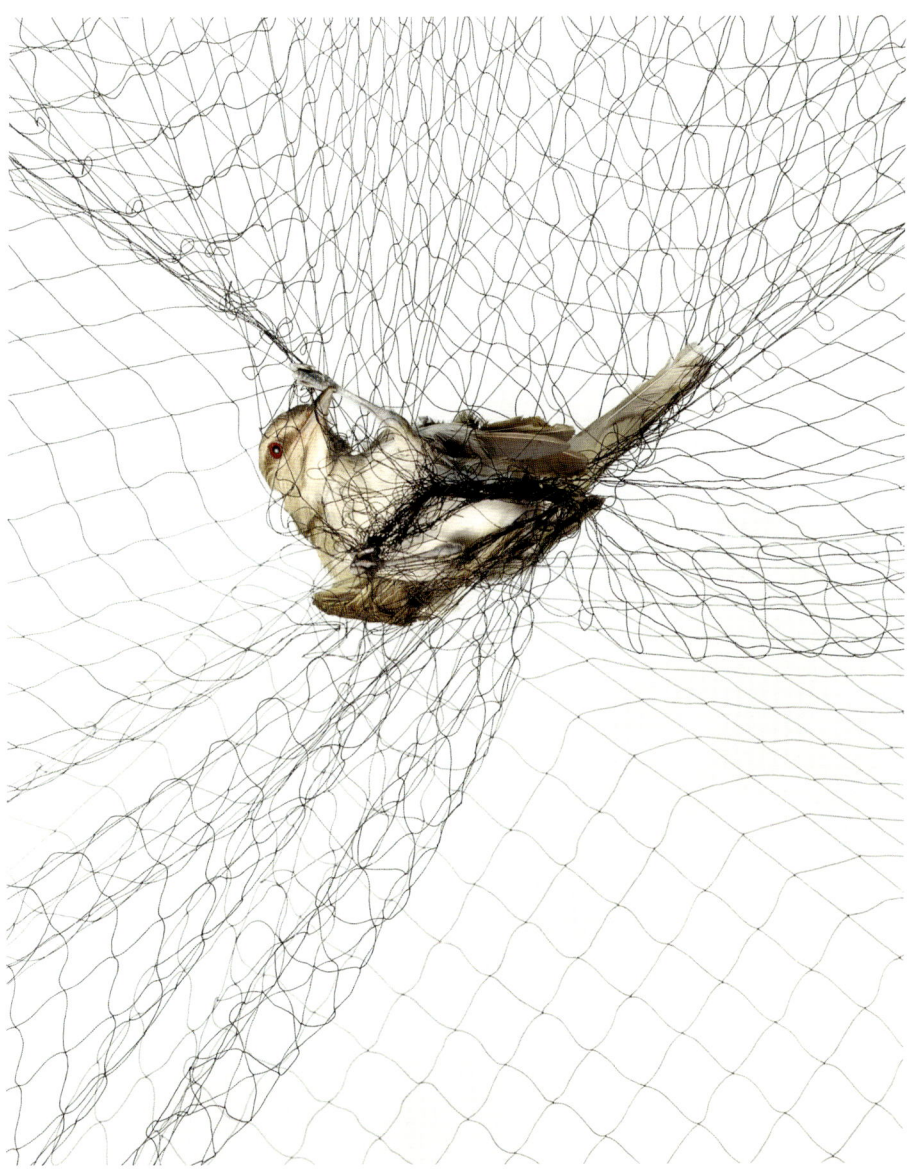

Black-whiskered Vireo
Vireo altiloquus

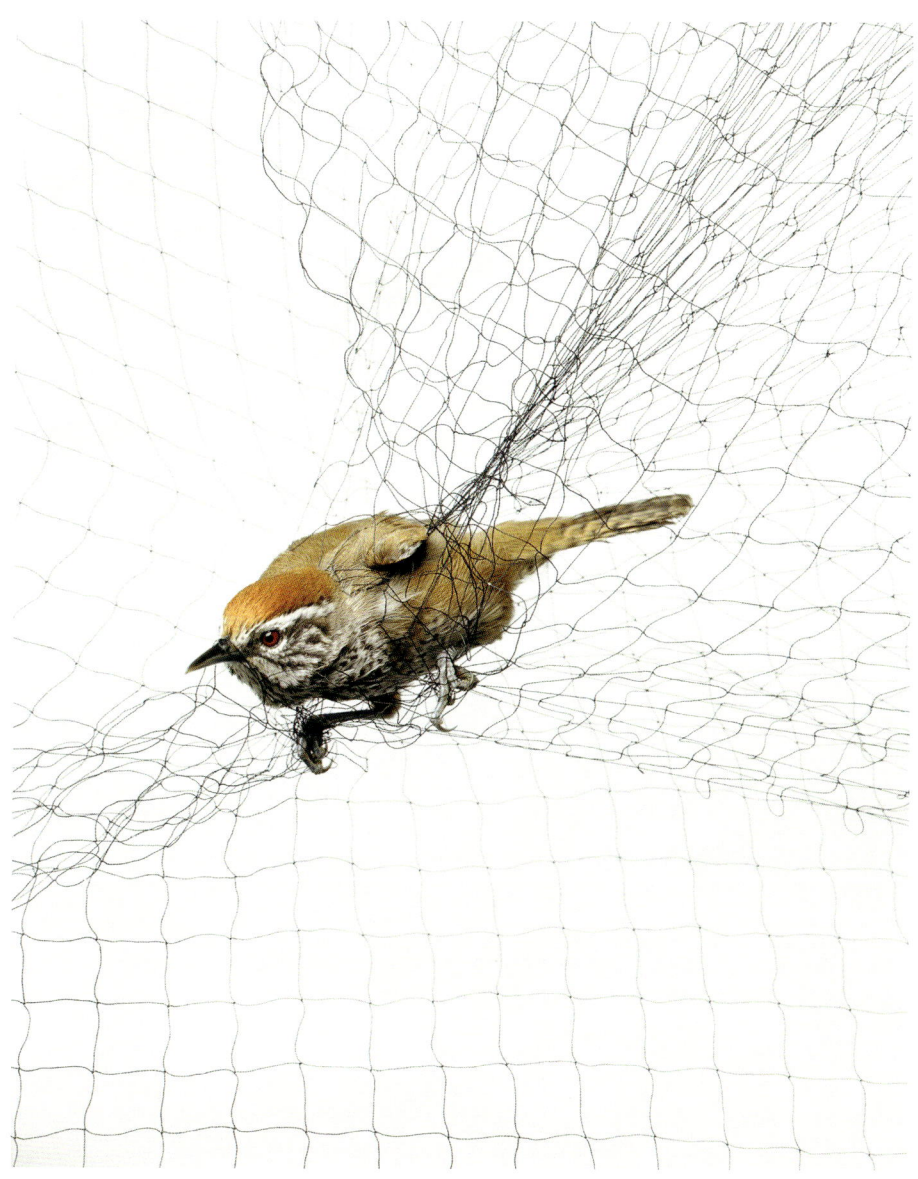

Spot-breasted Wren
Pheugopedius maculipectus

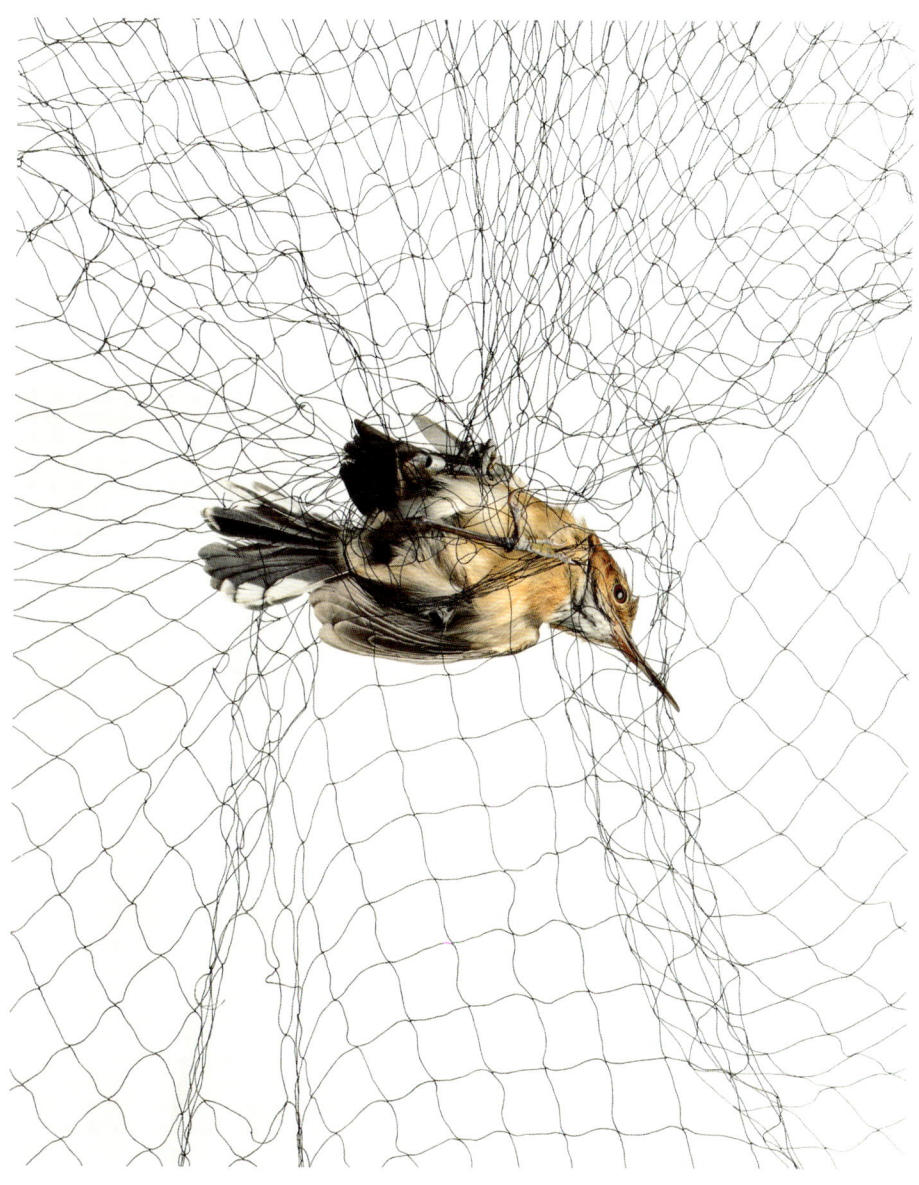

Long-billed Gnatwren
Ramphocaenus melanurus

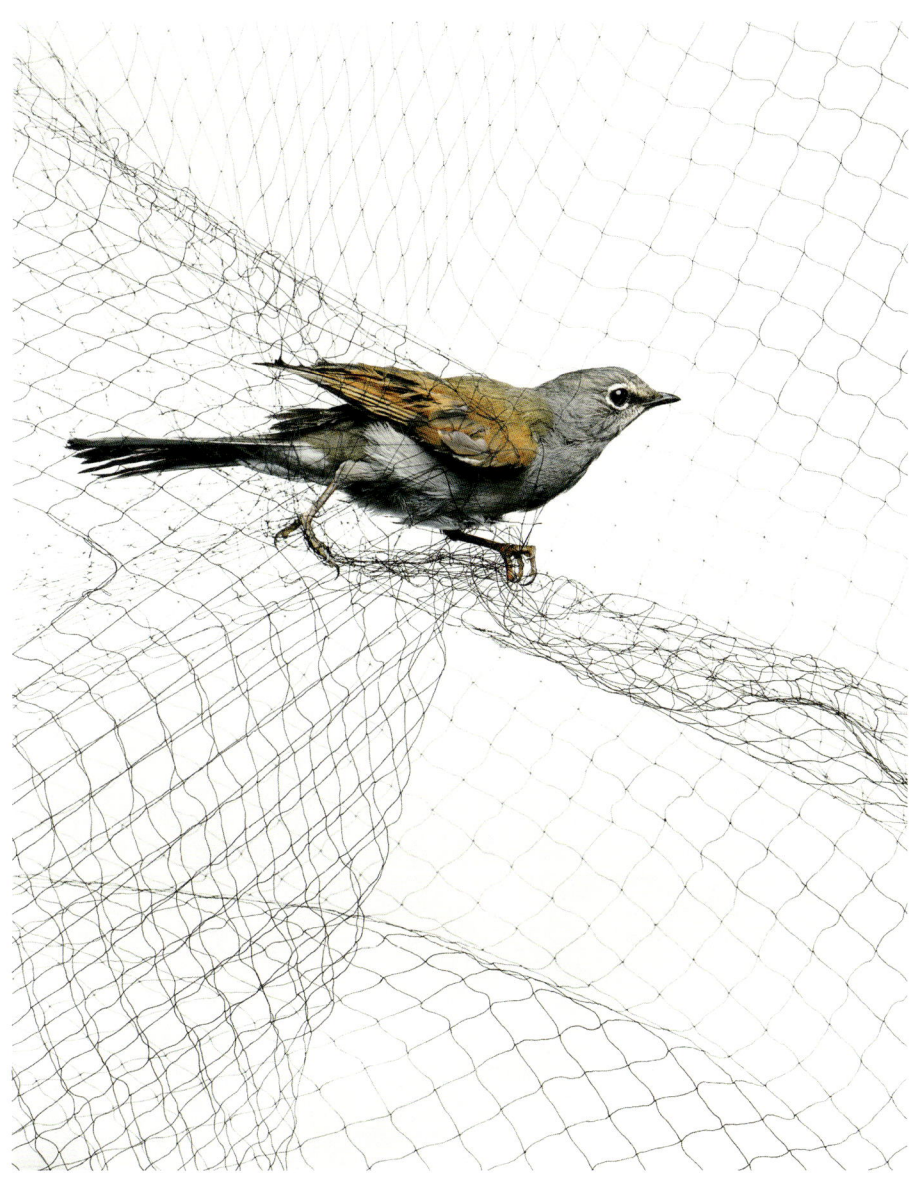

Brown-backed Solitaire
Myadestes occidentalis

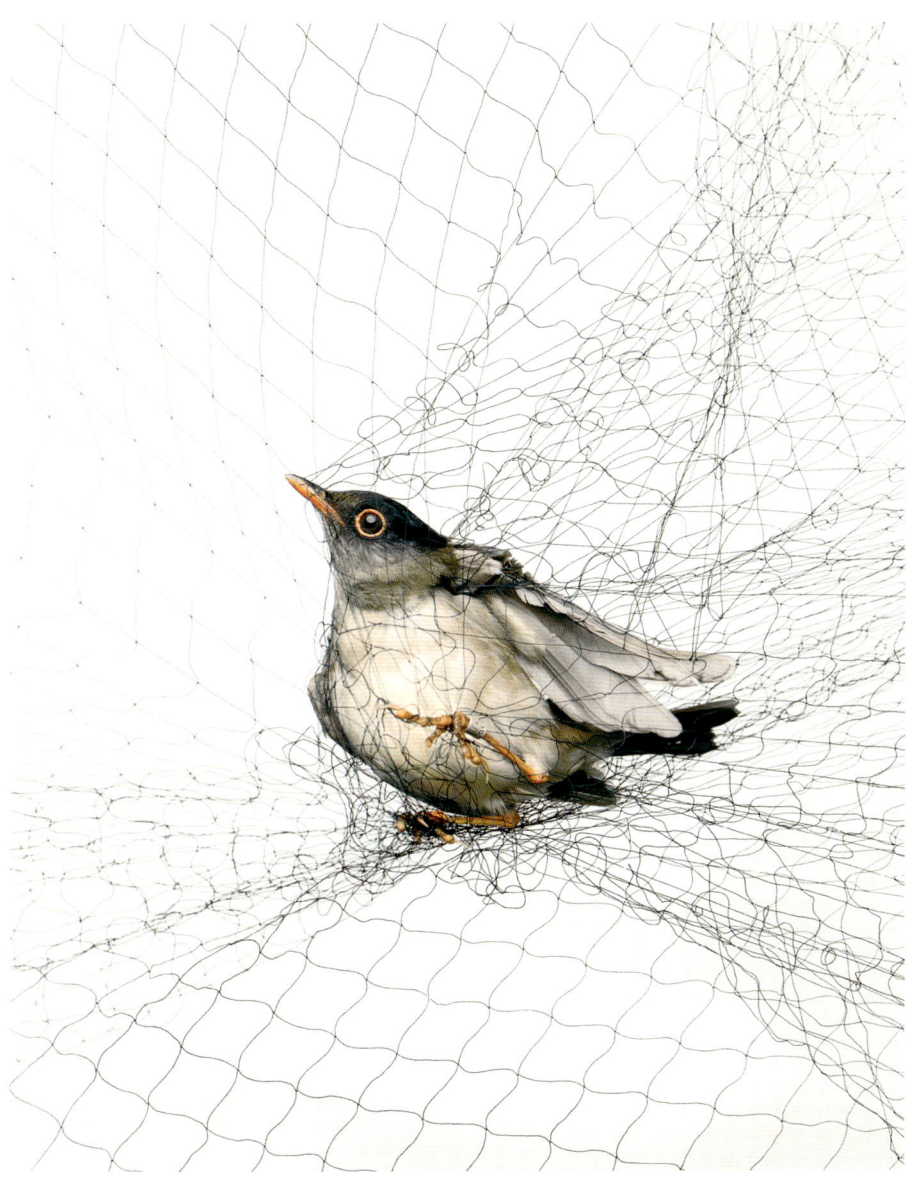

Black-headed Nightingale-thrush
Catharus mexicanus

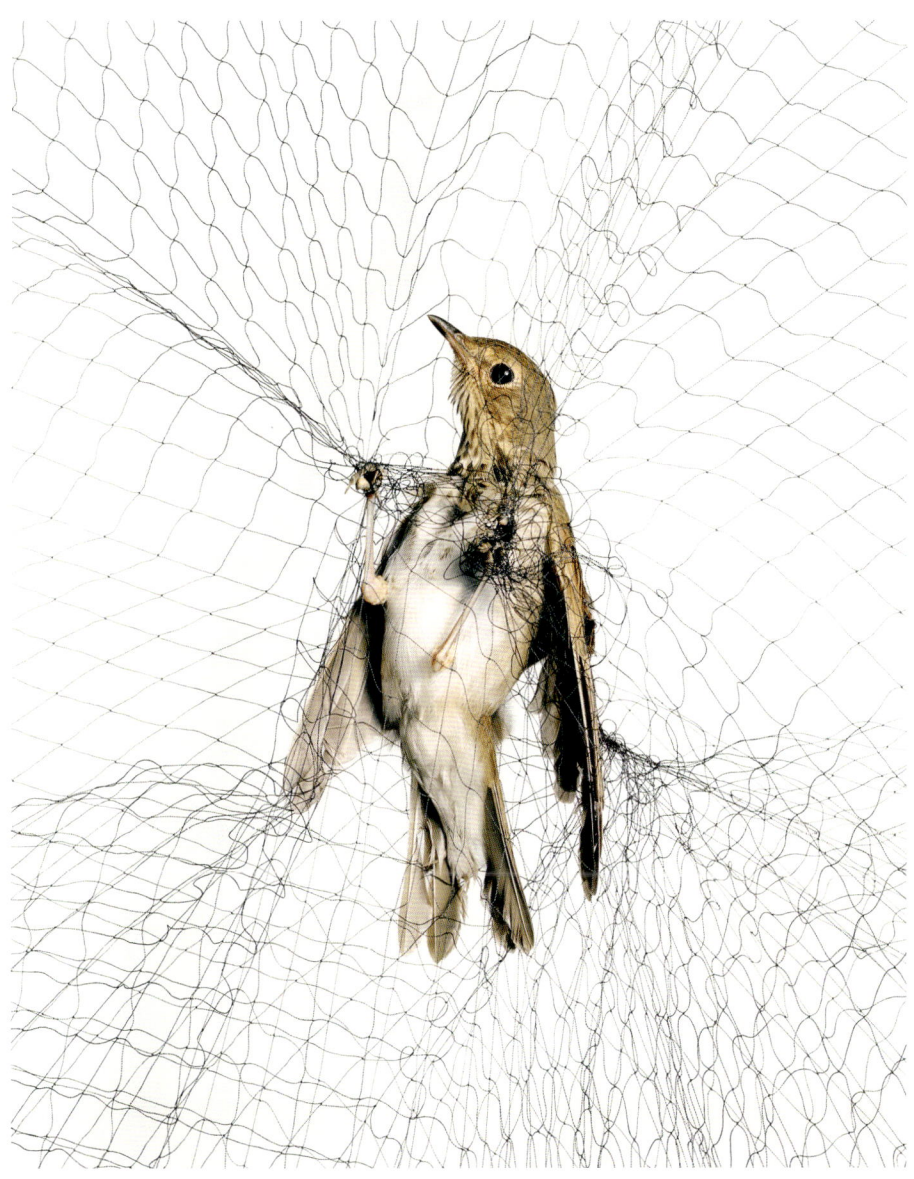

Swainson's Thrush
Catharus ustulatus

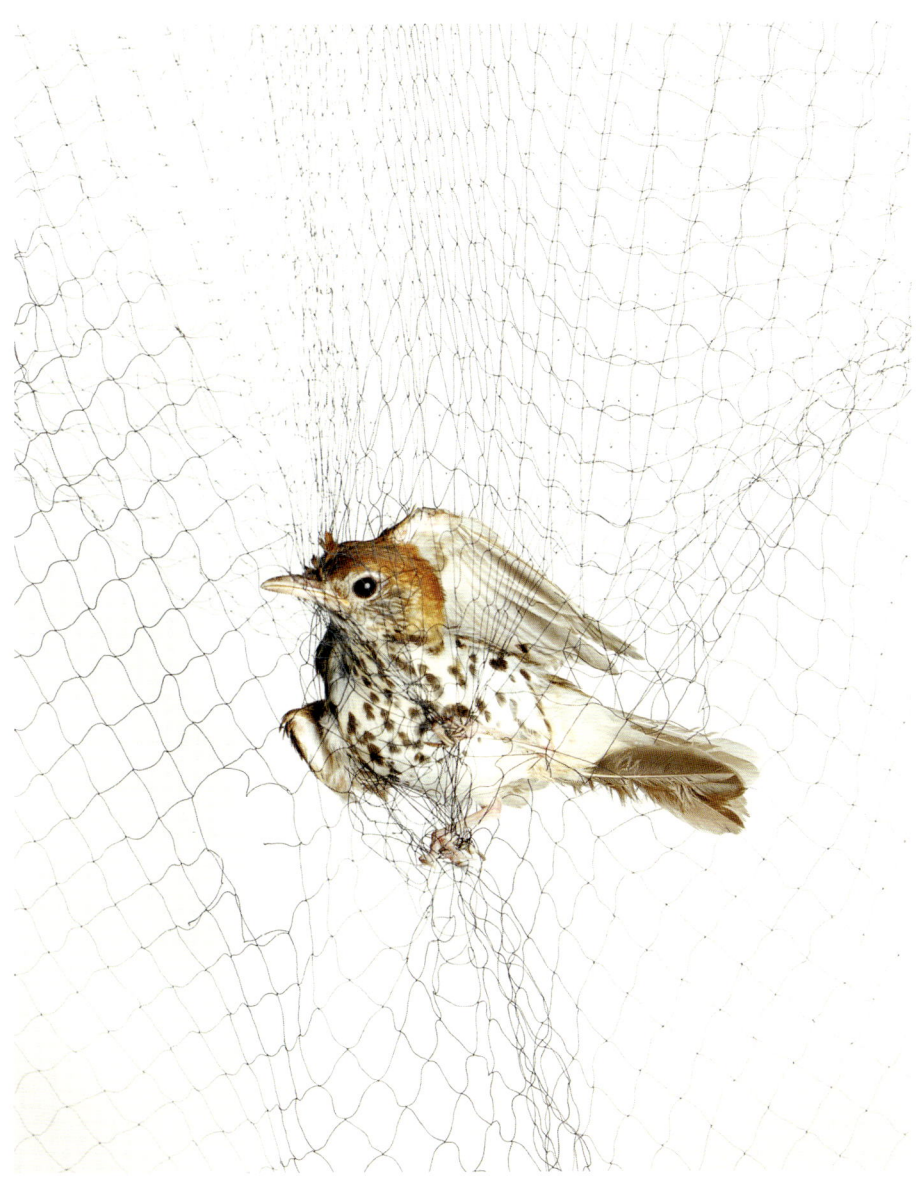

Wood Thrush
Hylocichla mustelina

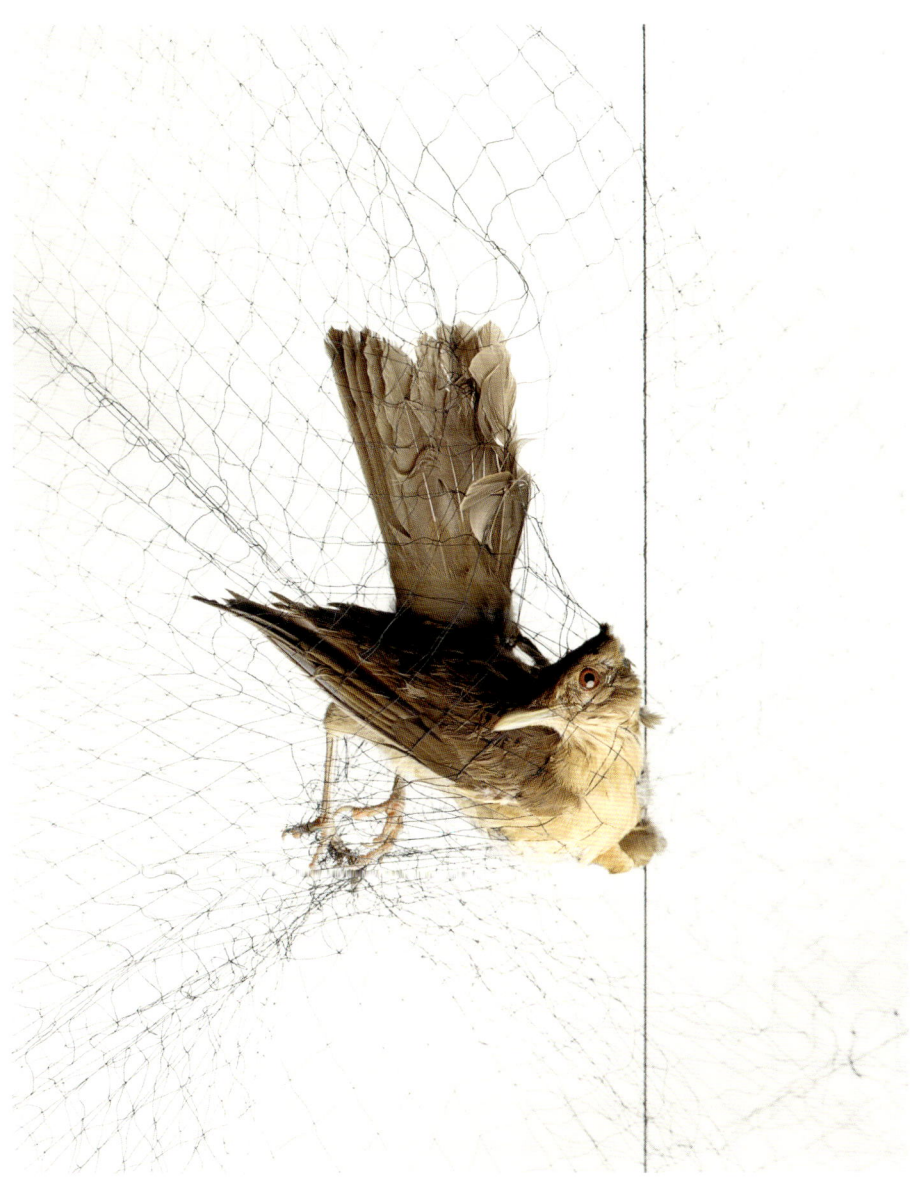

Clay-colored Thrush
Turdus grayi

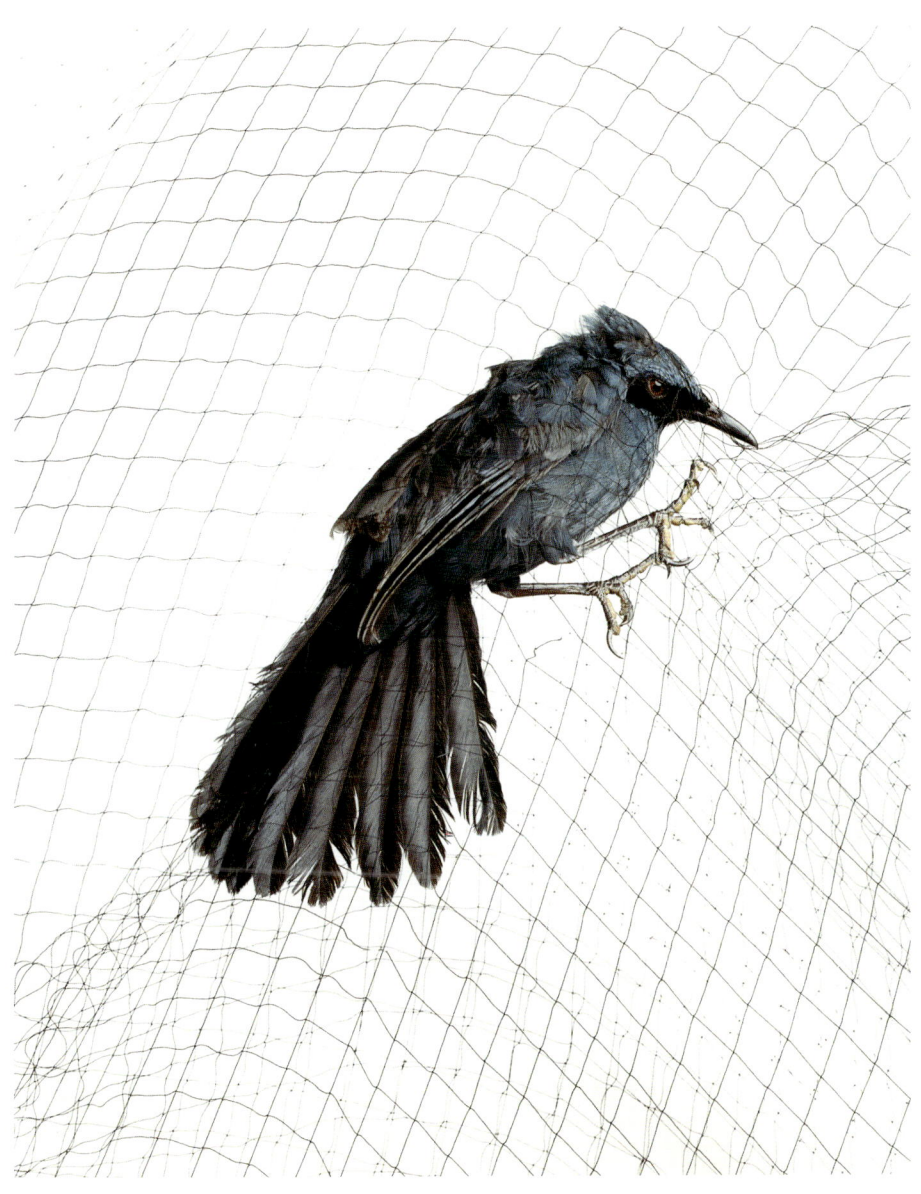

Blue Mockingbird
Melanotis caerulescens

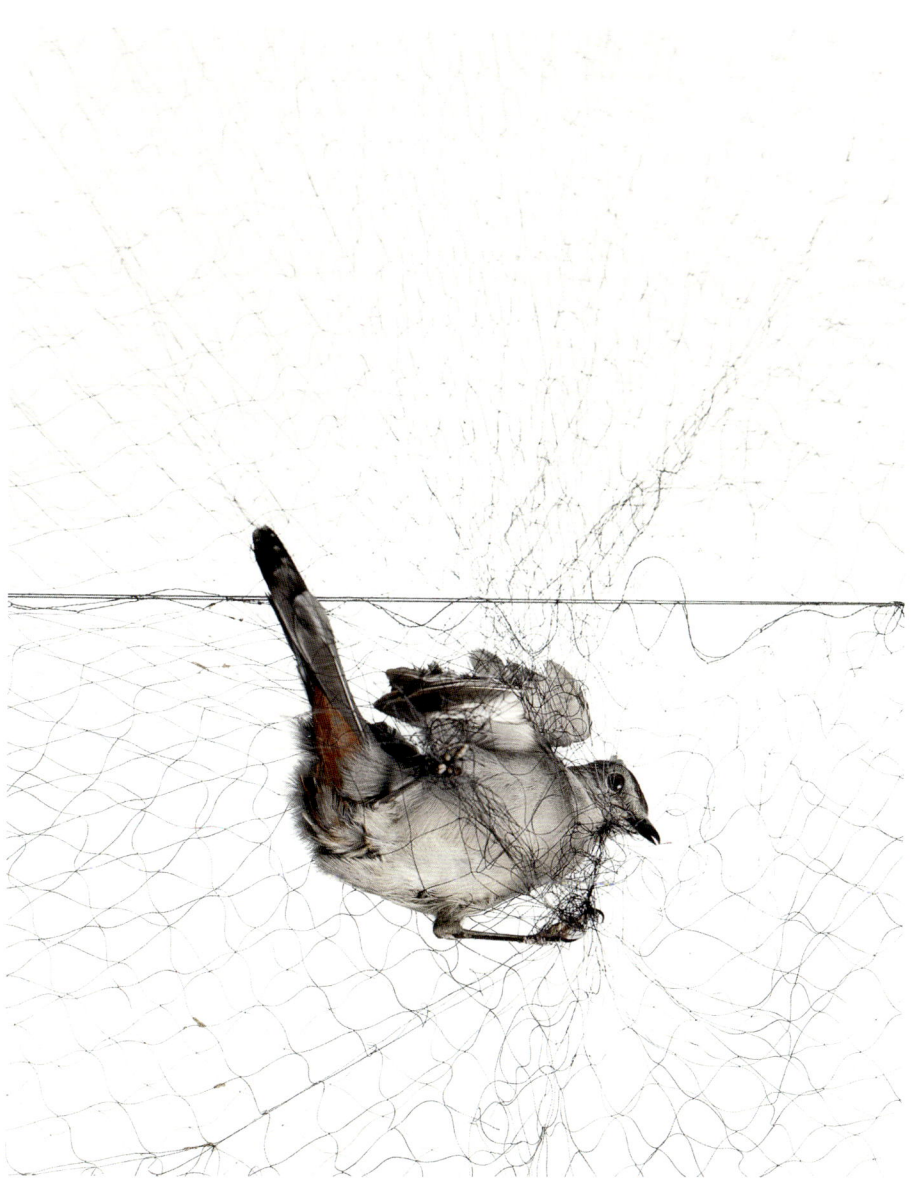

Gray Catbird
Dumetella carolinensis

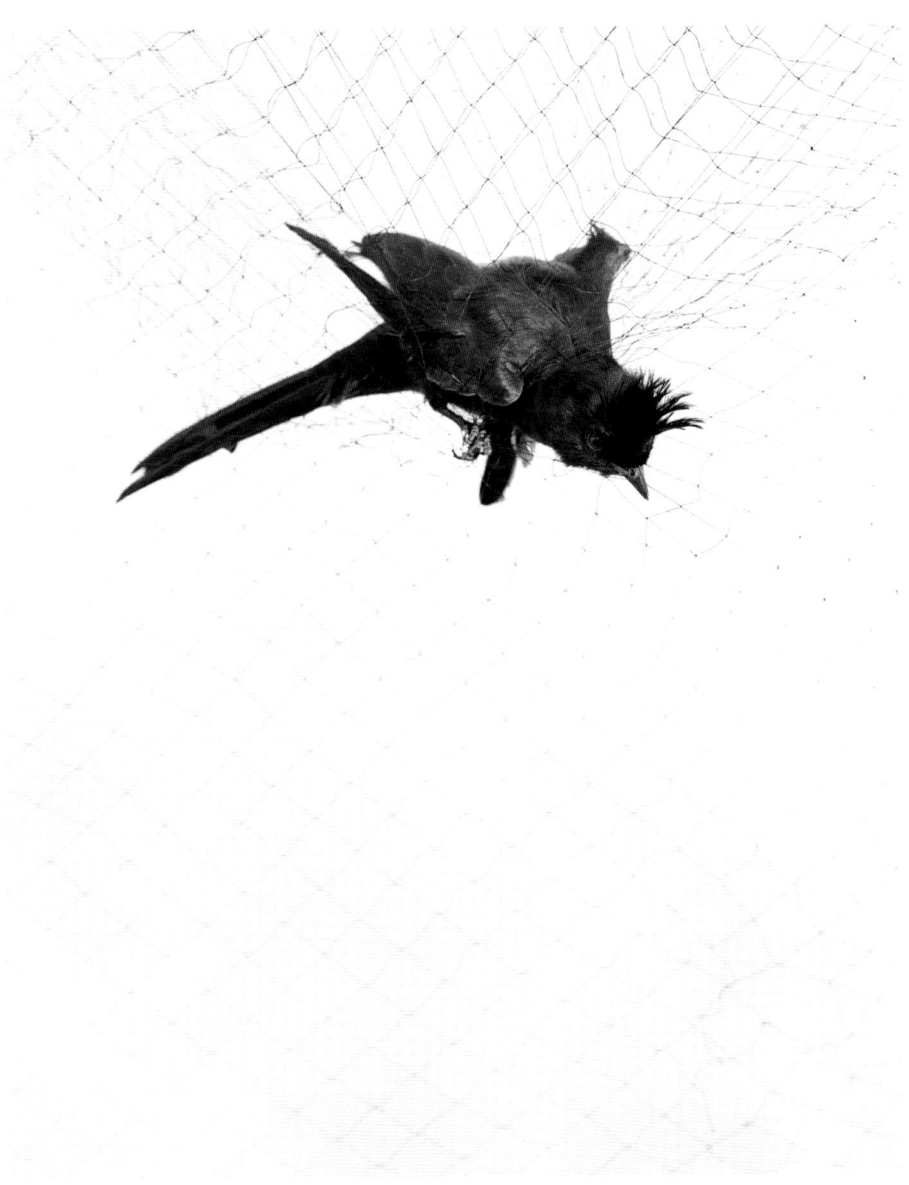

Phainopepla
Phainopepla nitens

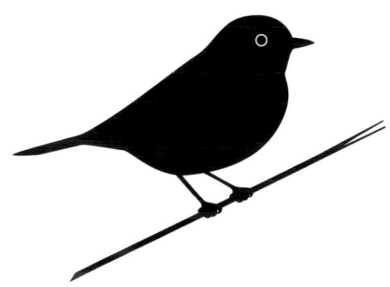

PASSERINES
Oscines
New World warblers

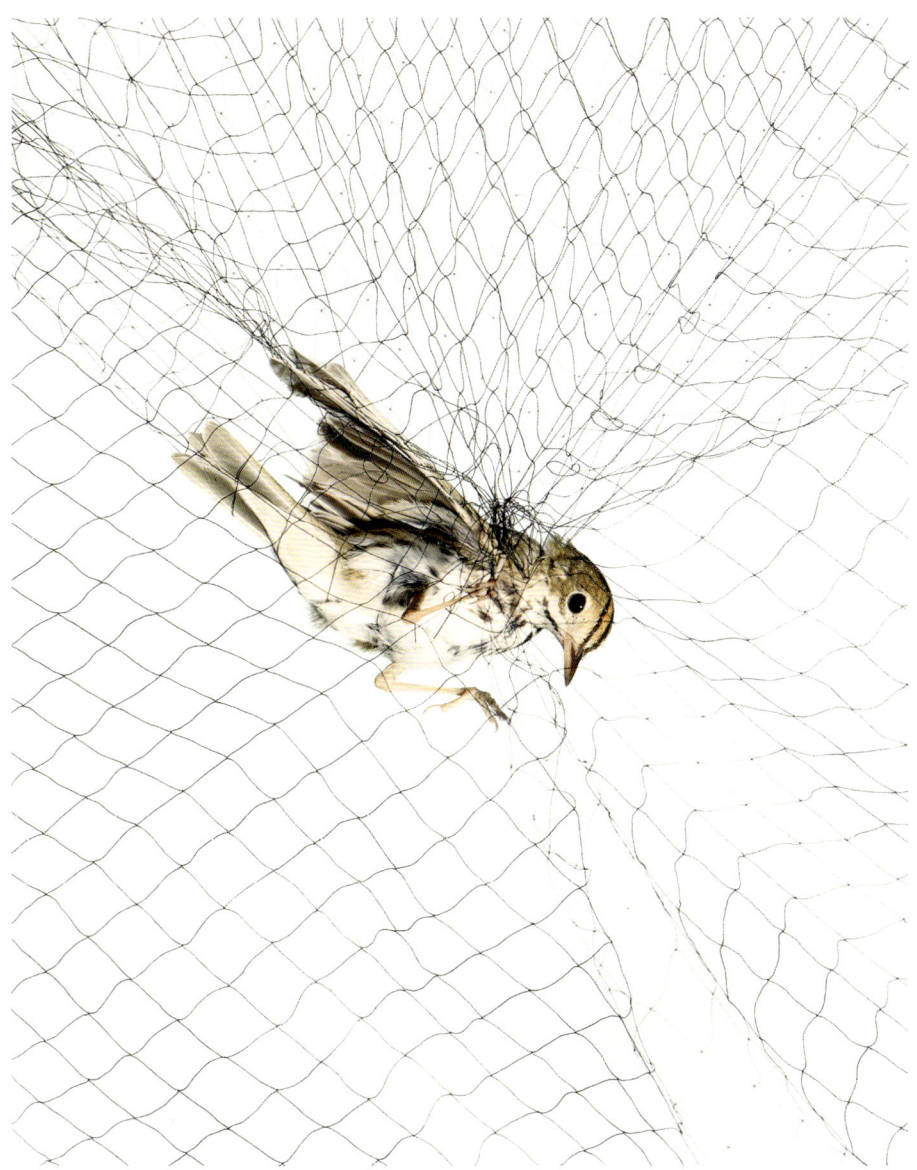

Ovenbird
Seiurus aurocapilla

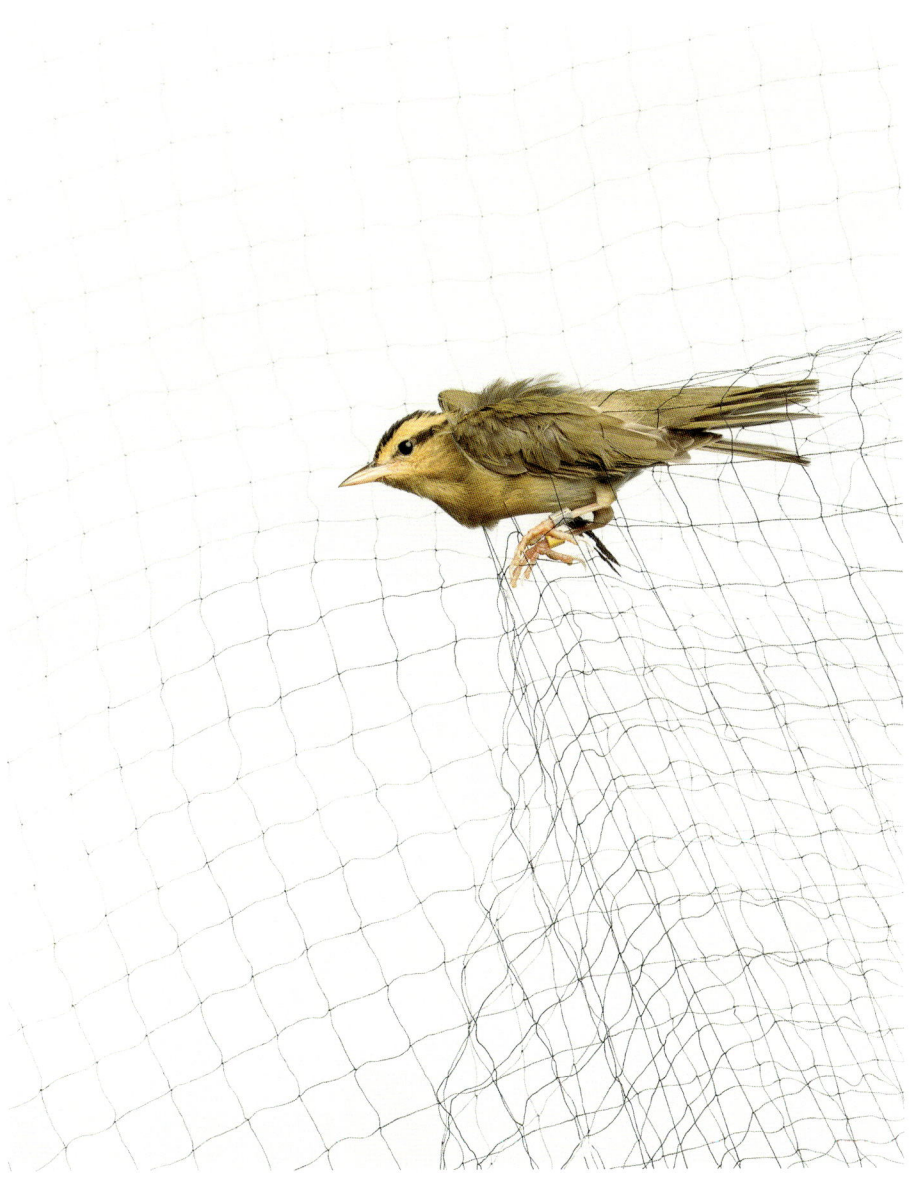

Worm-eating Warbler
Helmitheros vermivorum

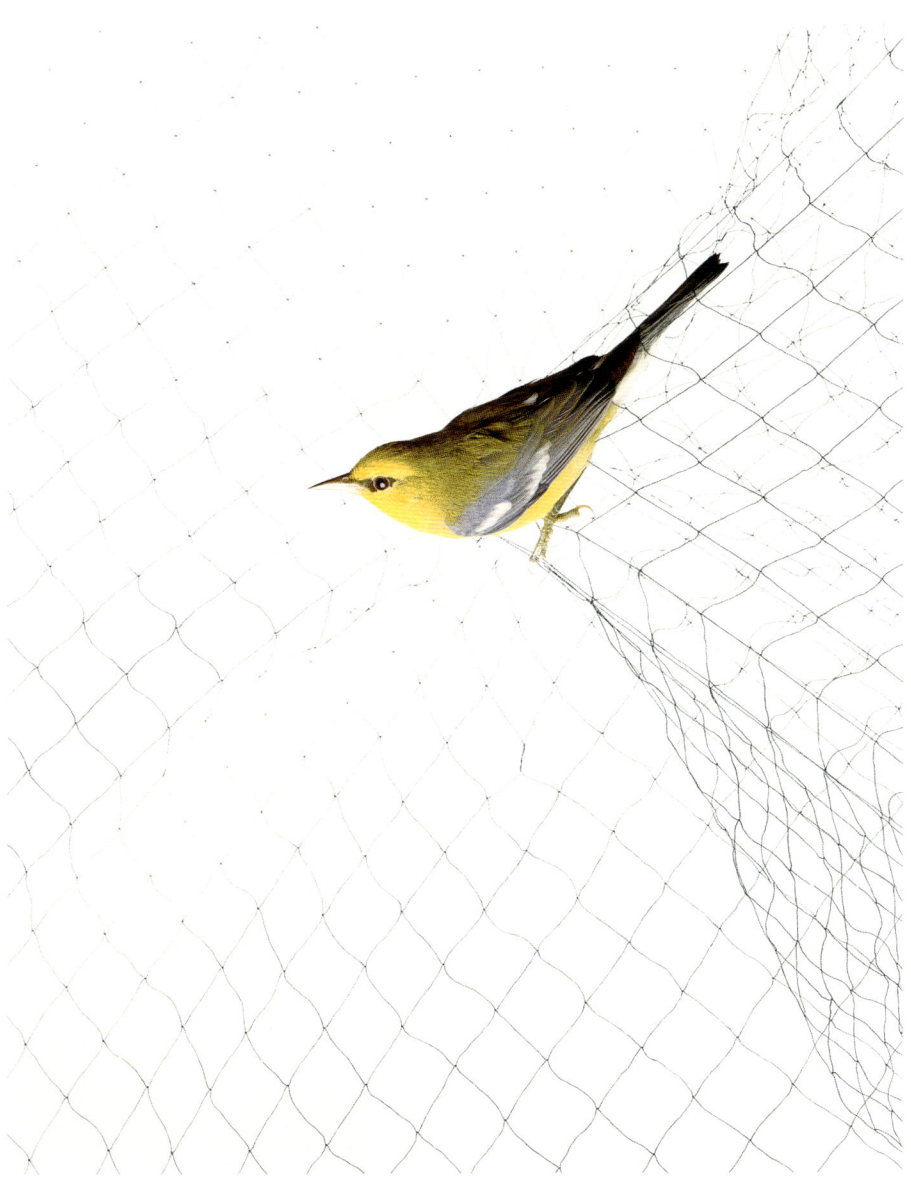

Blue-winged Warbler
Vermivora cyanoptera

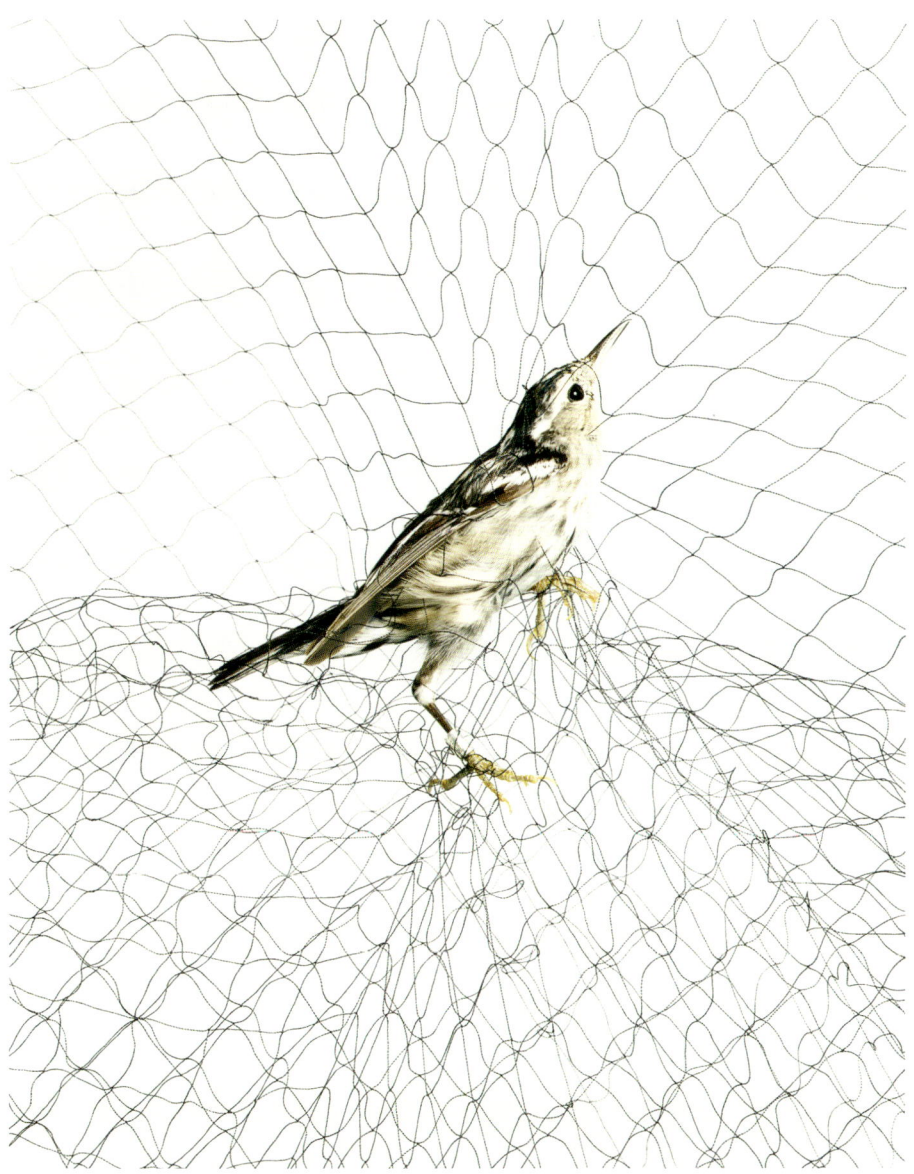

Black-and-white Warbler
Mniotilta varia

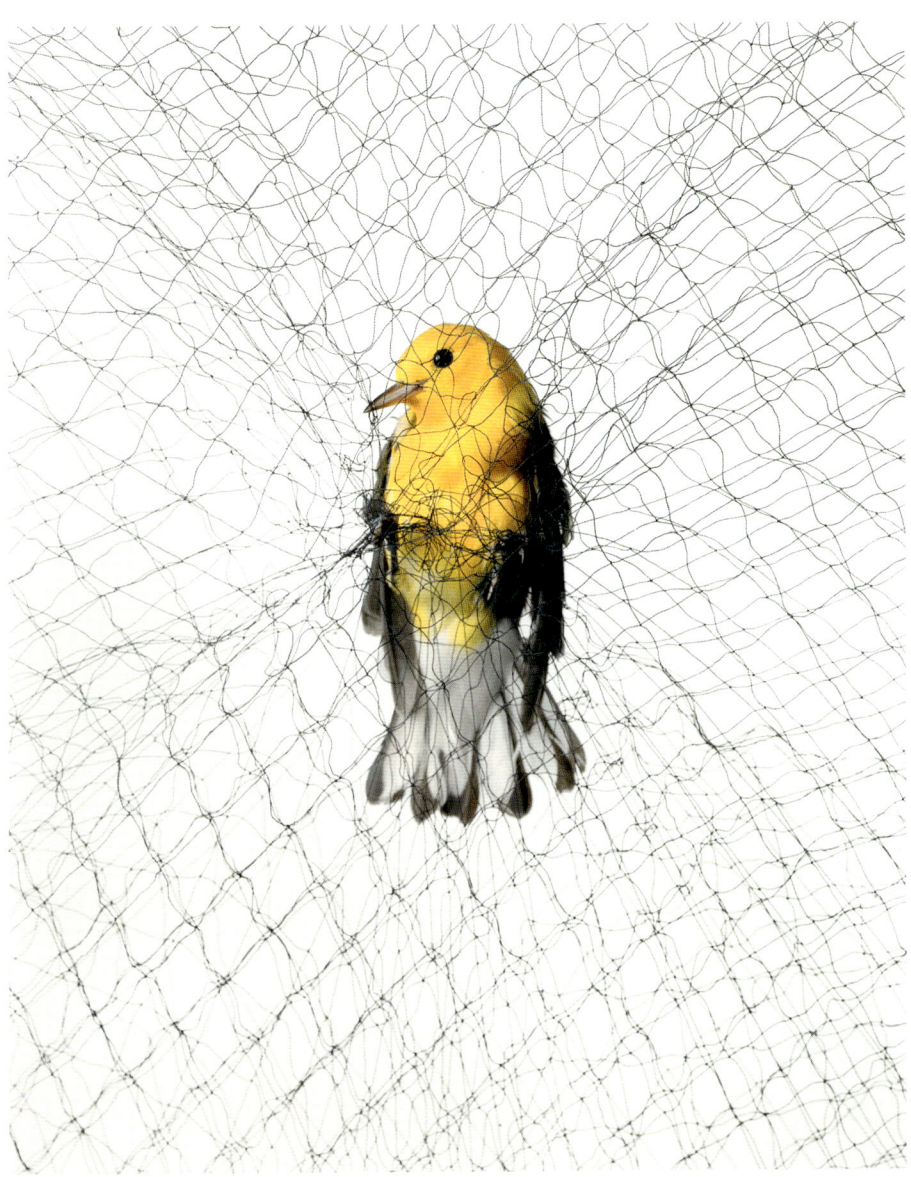

Prothonotary Warbler
Protonotaria citrea

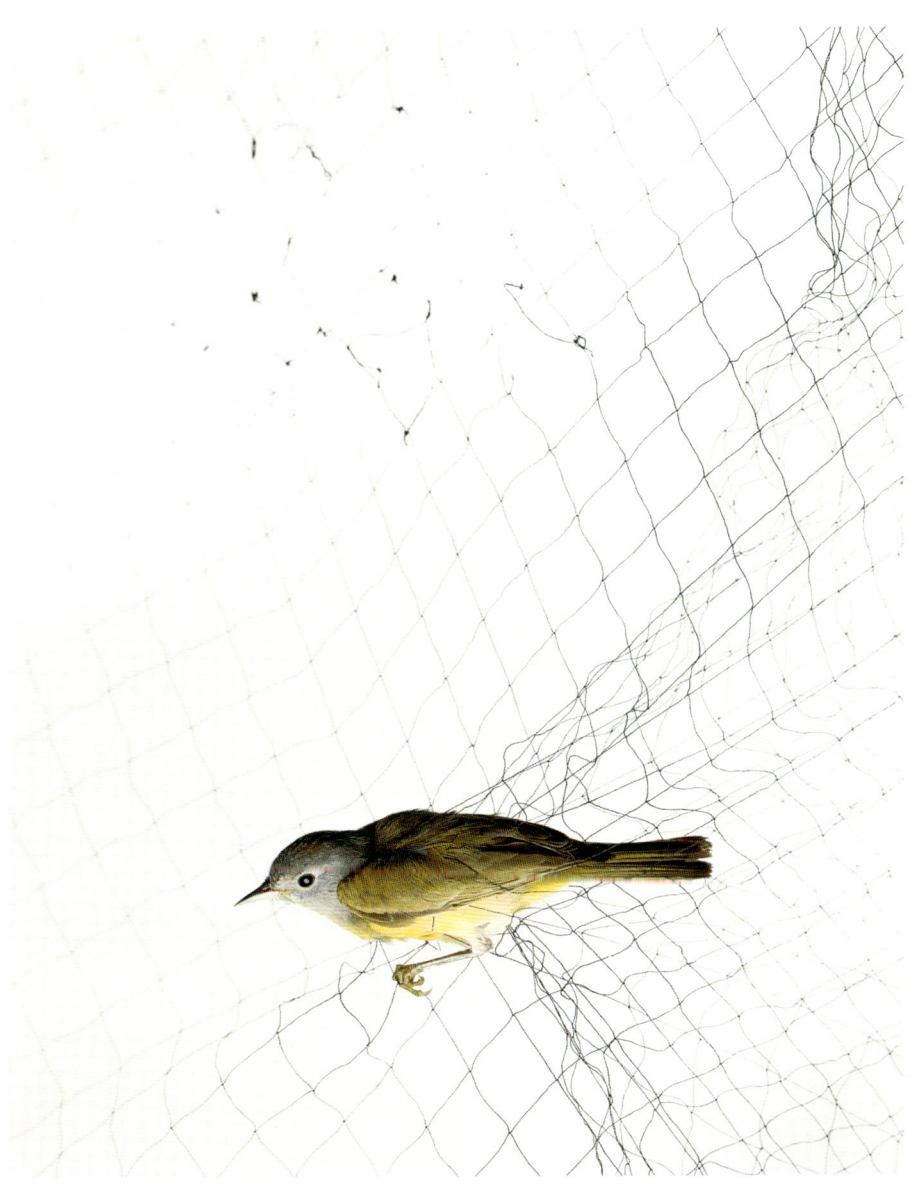

Nashville Warbler
Oreothlypis ruficapilla

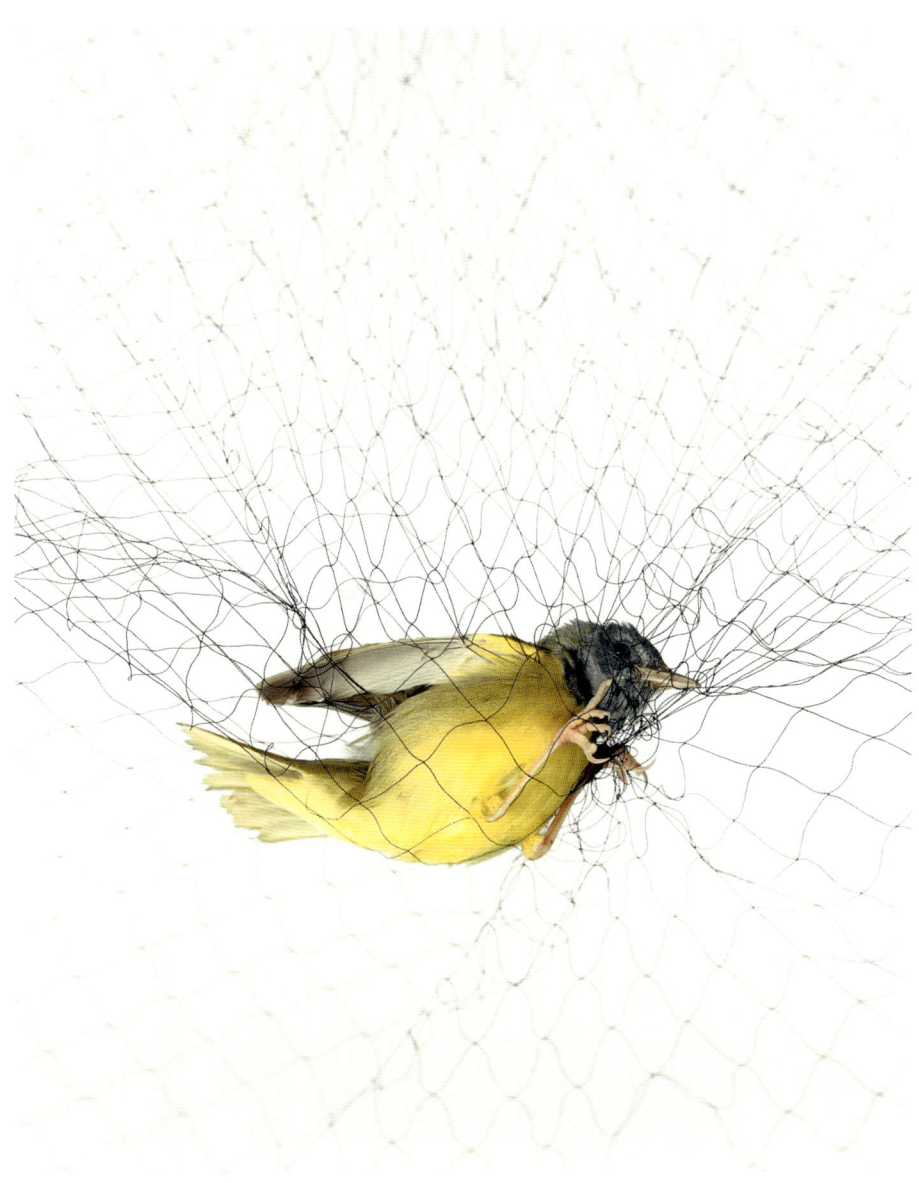

Mourning Warbler
Geothlypis philadelphia

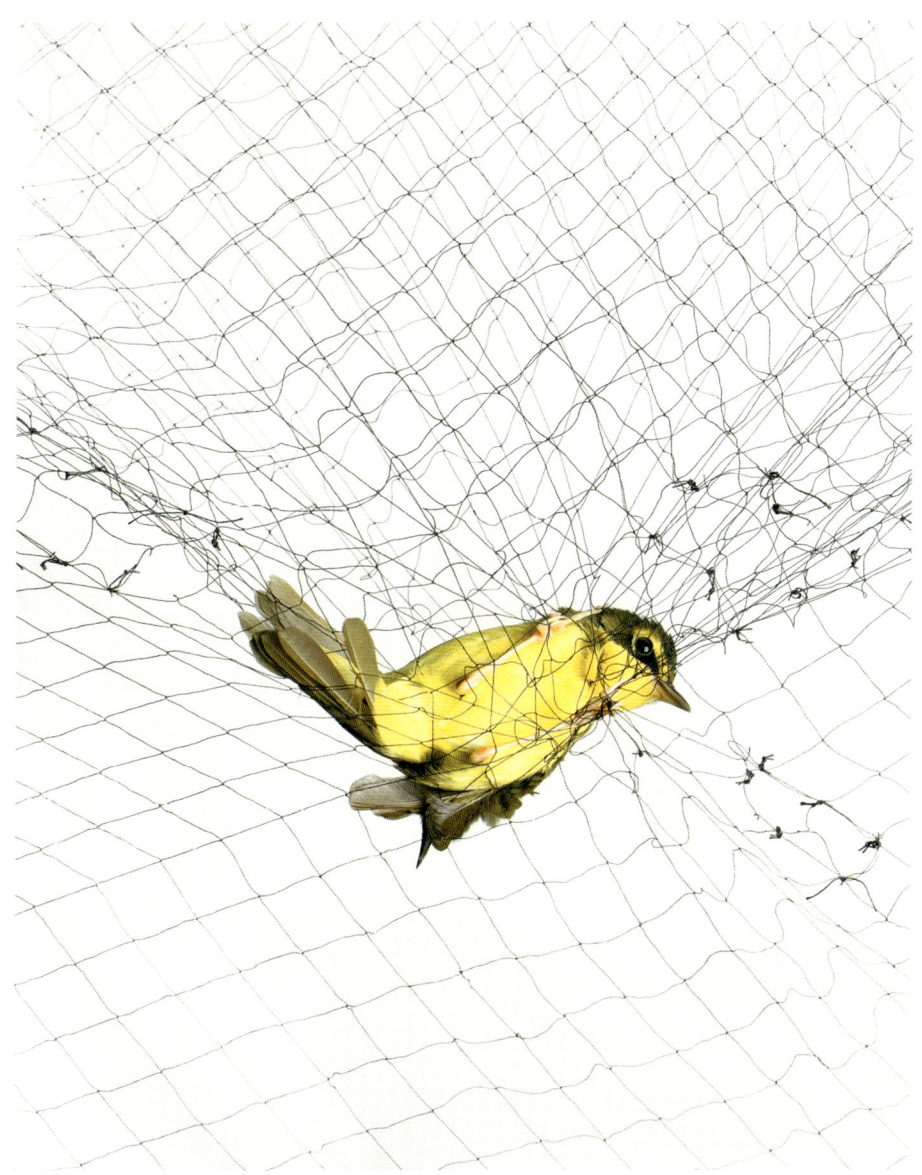

Kentucky Warbler
Geothlypis formosa

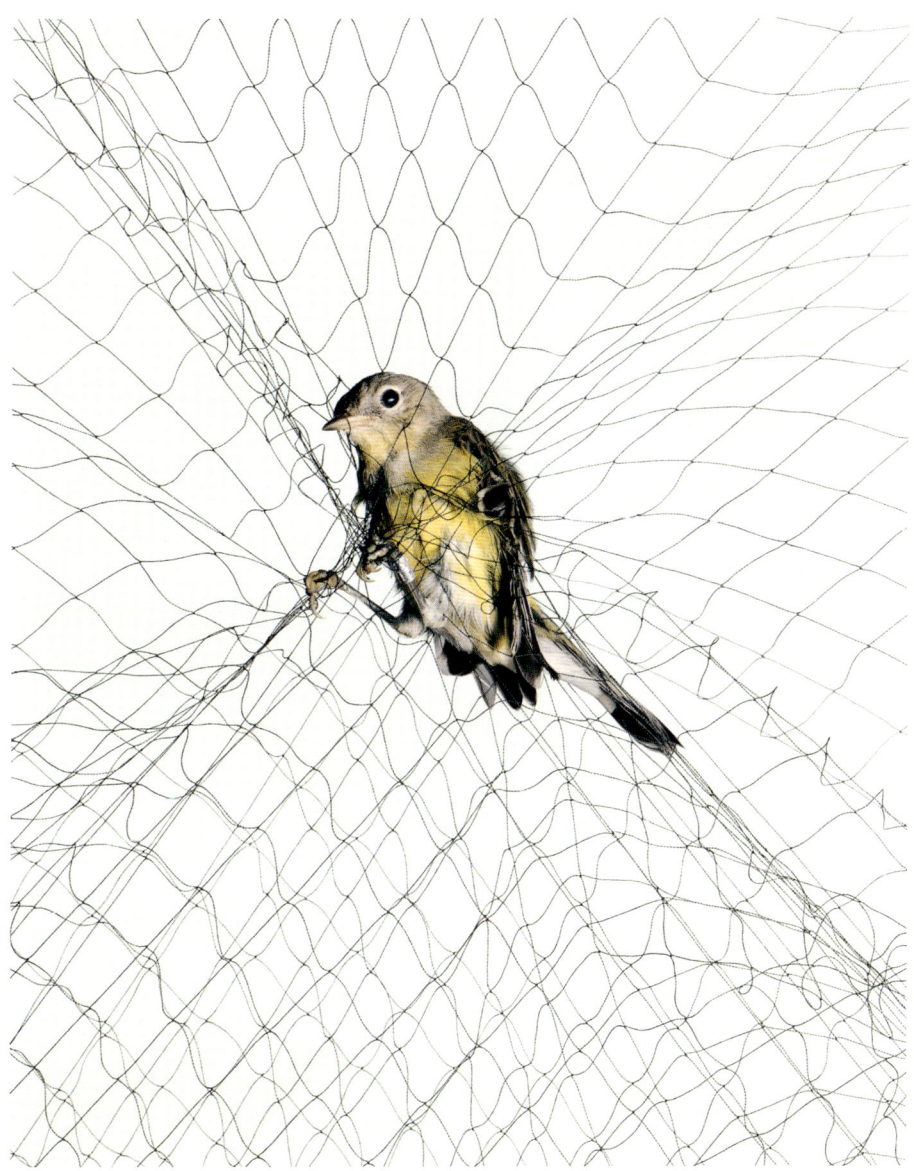

Magnolia Warbler
Setophaga magnolia

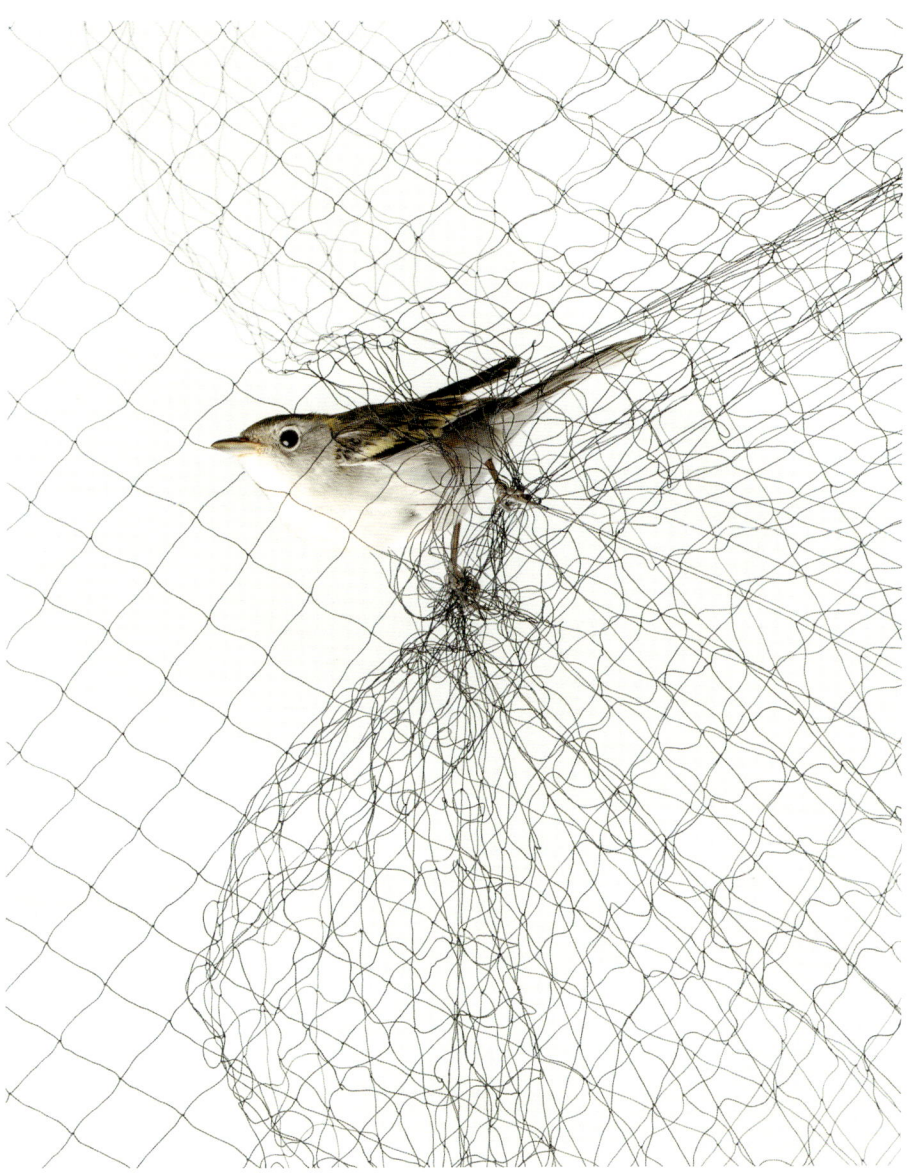

Chestnut-sided Warbler
Setophaga pensylvanica

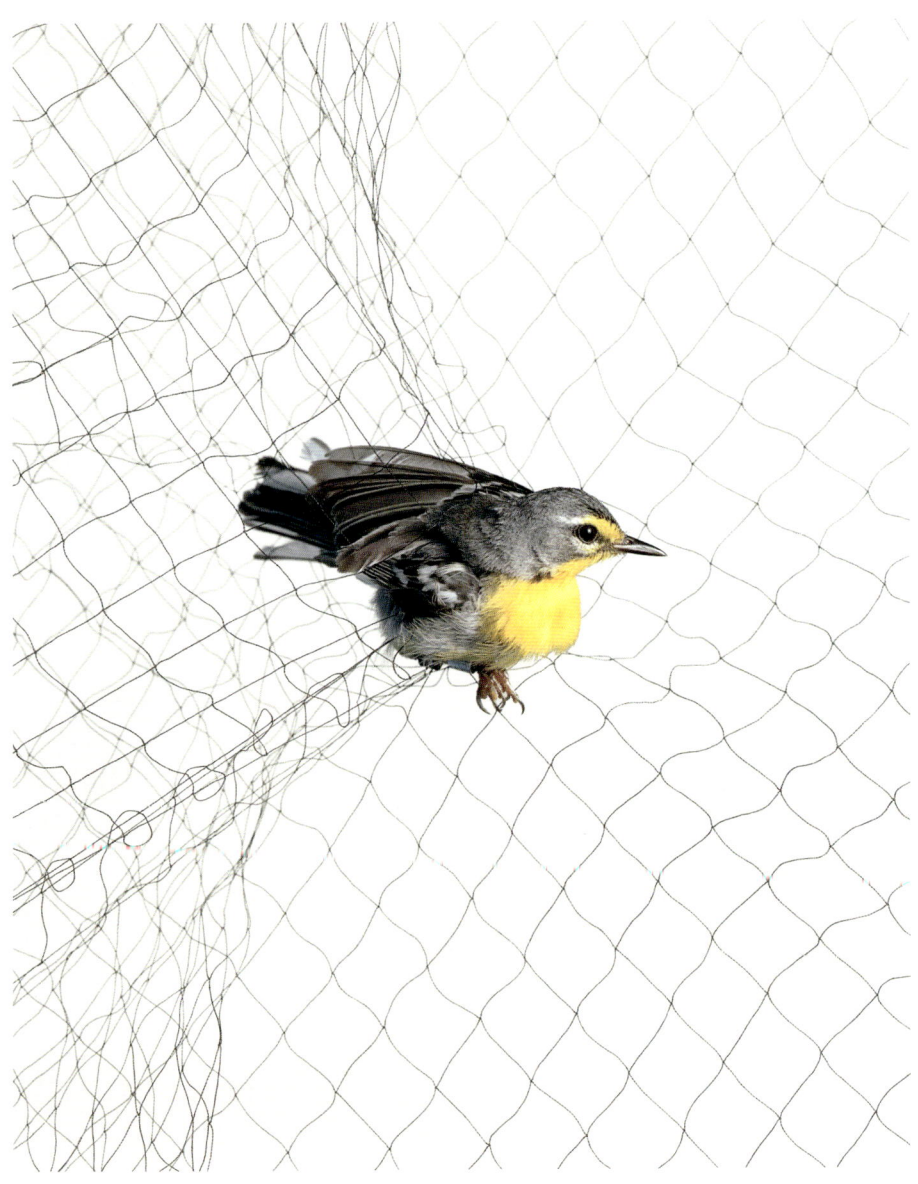

Adelaide's Warbler
Setophaga adelaidae

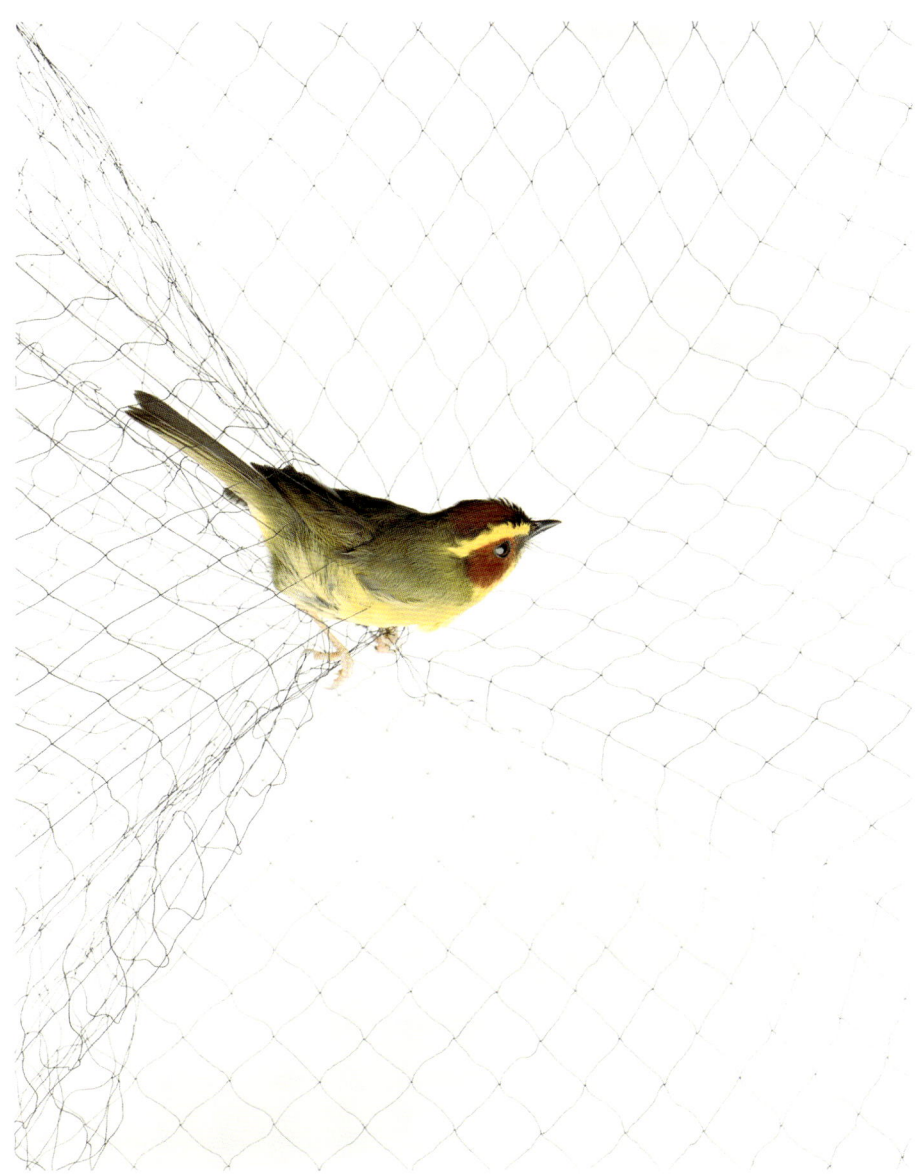

Golden-browed Warbler
Basileuterus belli

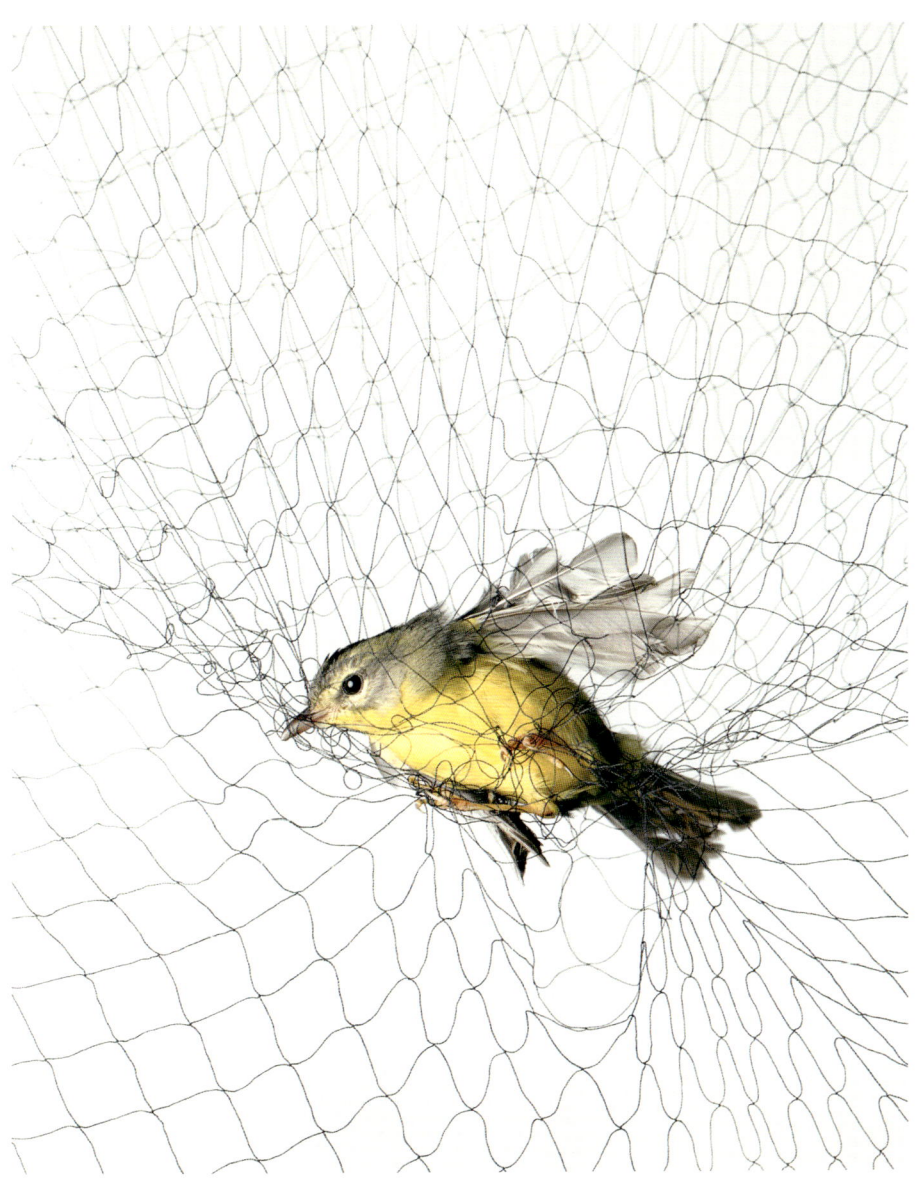

Golden-crowned Warbler
Basileuterus culicivorus

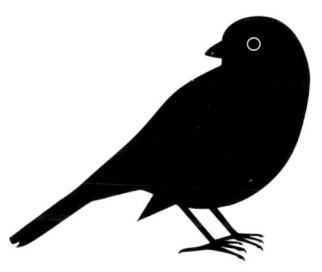

PASSERINES
Oscines
seedeaters to oropendolas

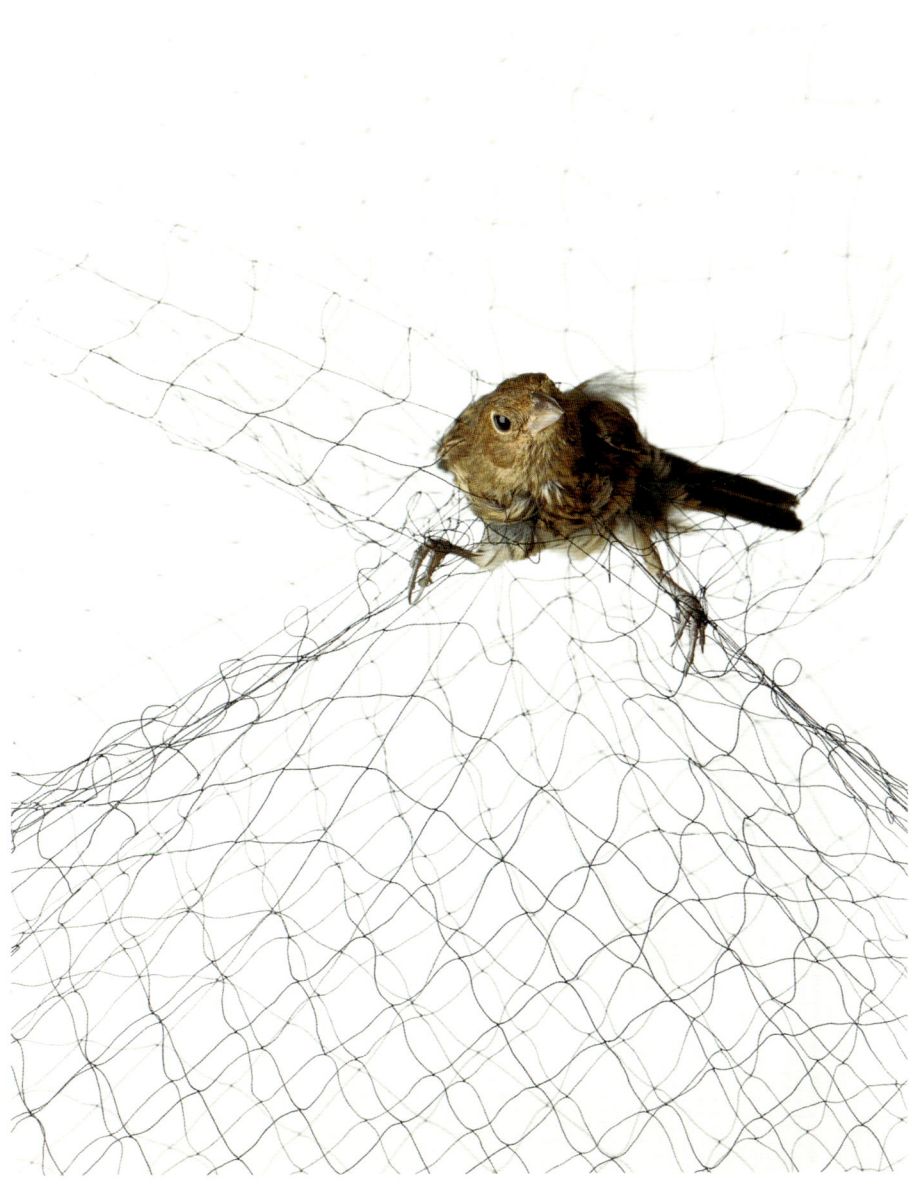

Variable Seedeater
Sporophila corvina

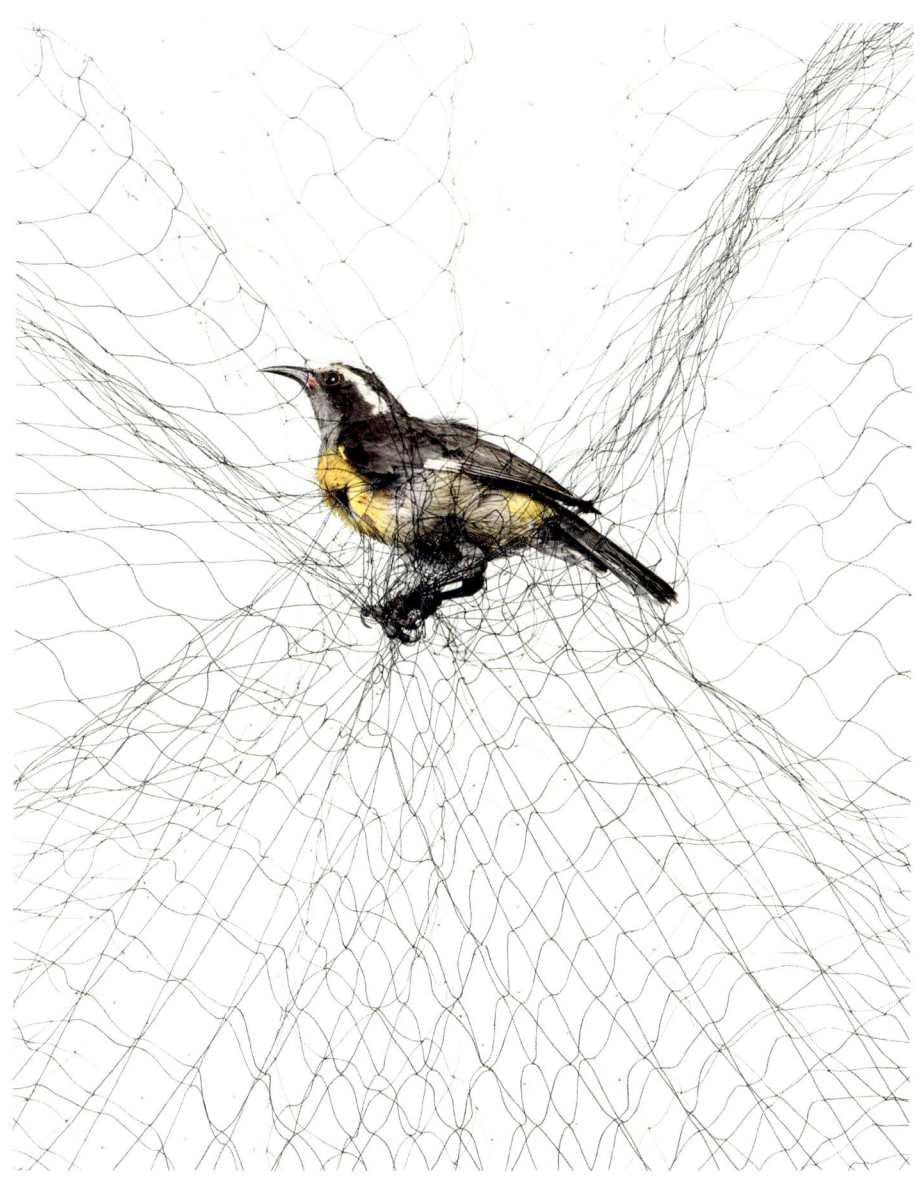

Bananaquit
Coereba flaveola

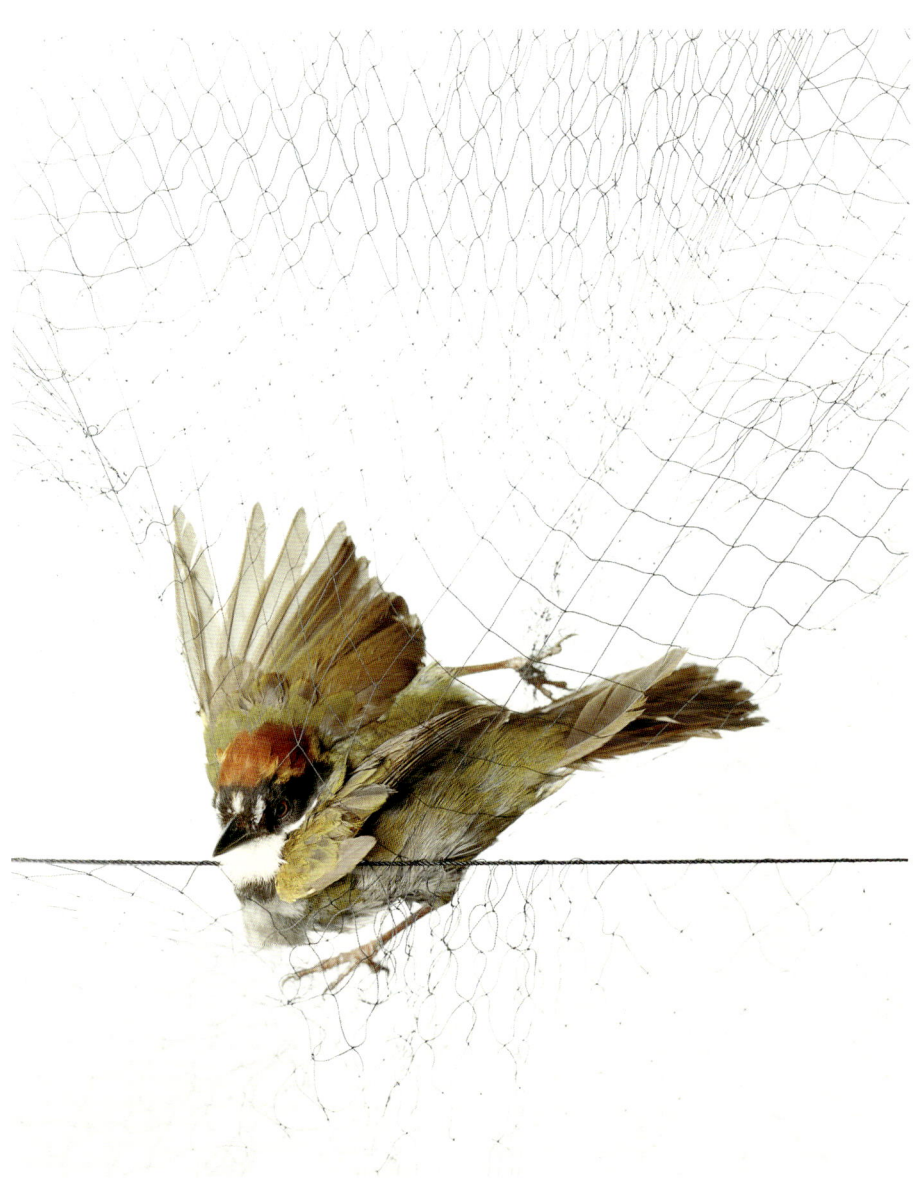

Chestnut-capped Brush-finch
Arremon brunneinucha

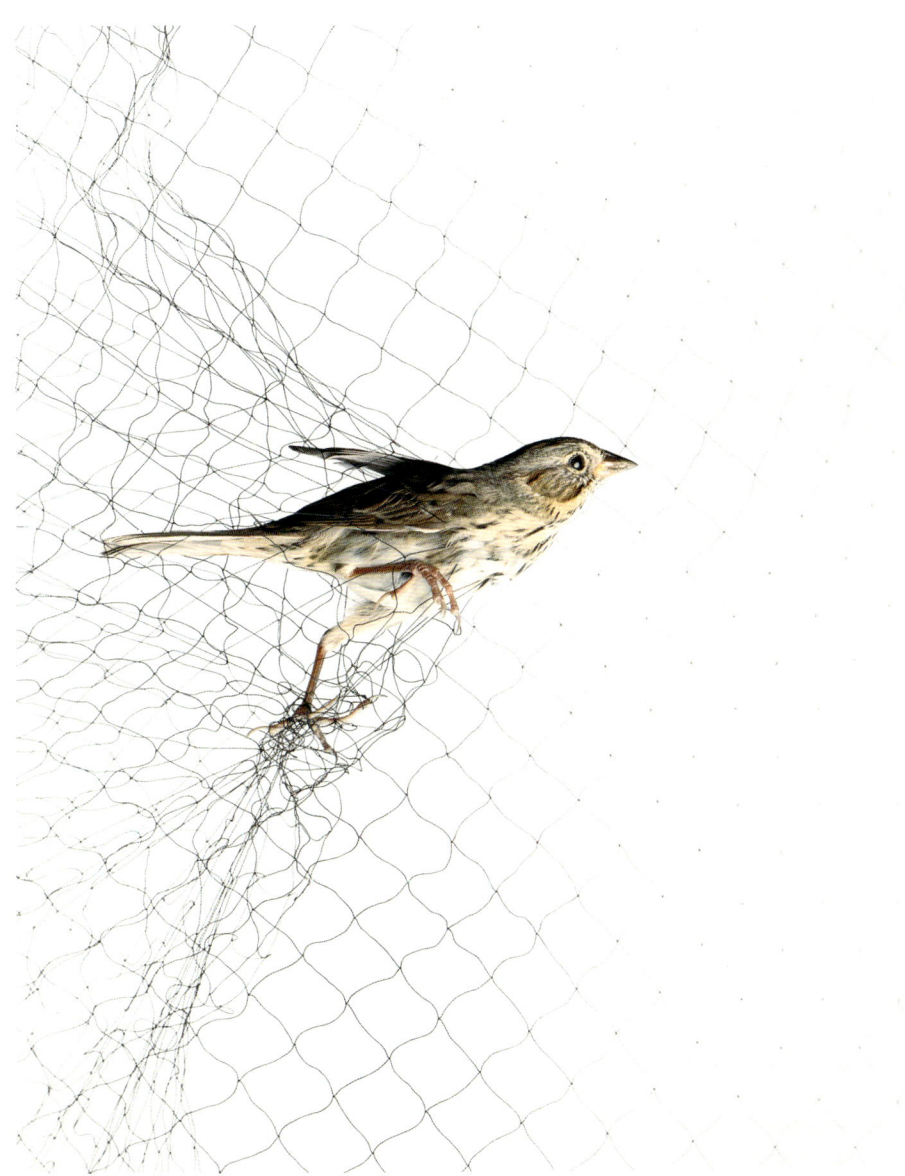

Lincoln's Sparrow
Melospiza lincolnii

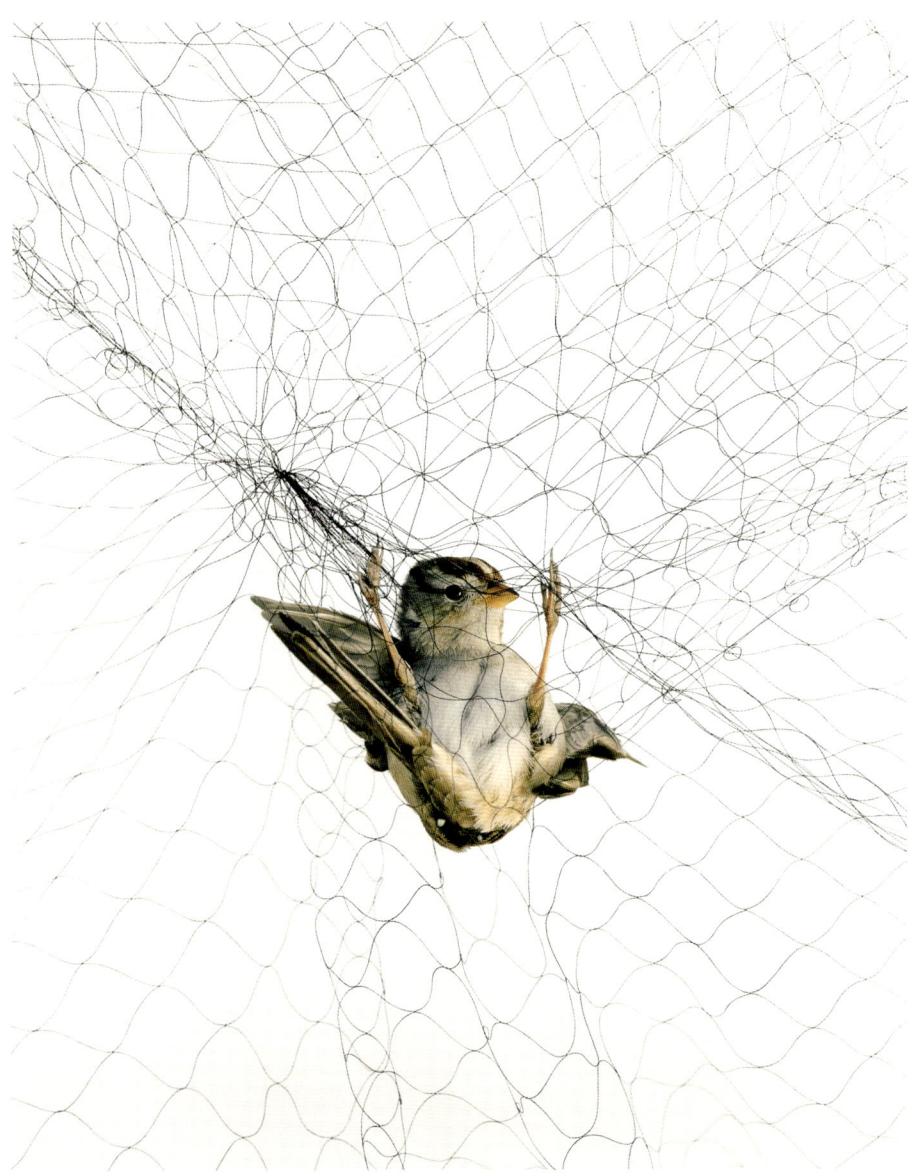

White-crowned Sparrow
Zonotrichia leucophrys

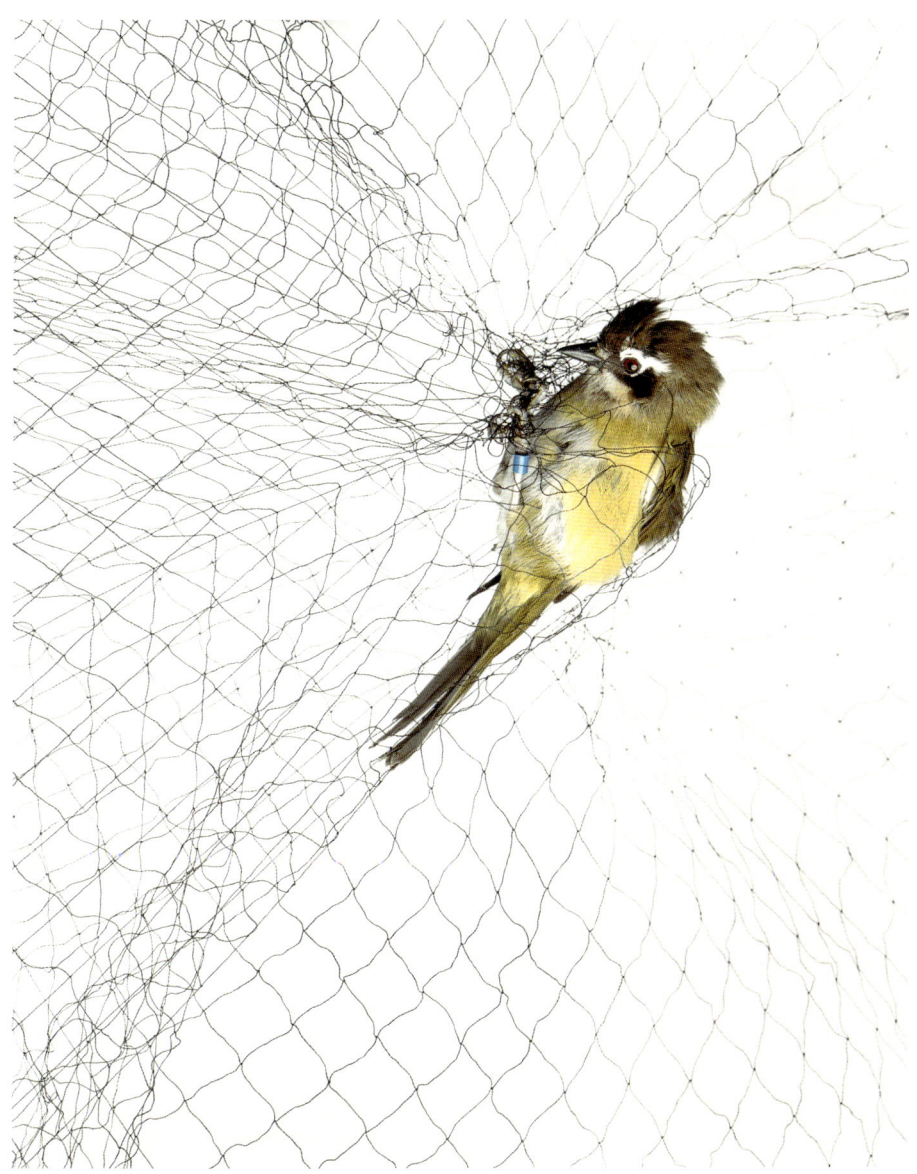

Common Chlorospingus
Chlorospingus flavopectus

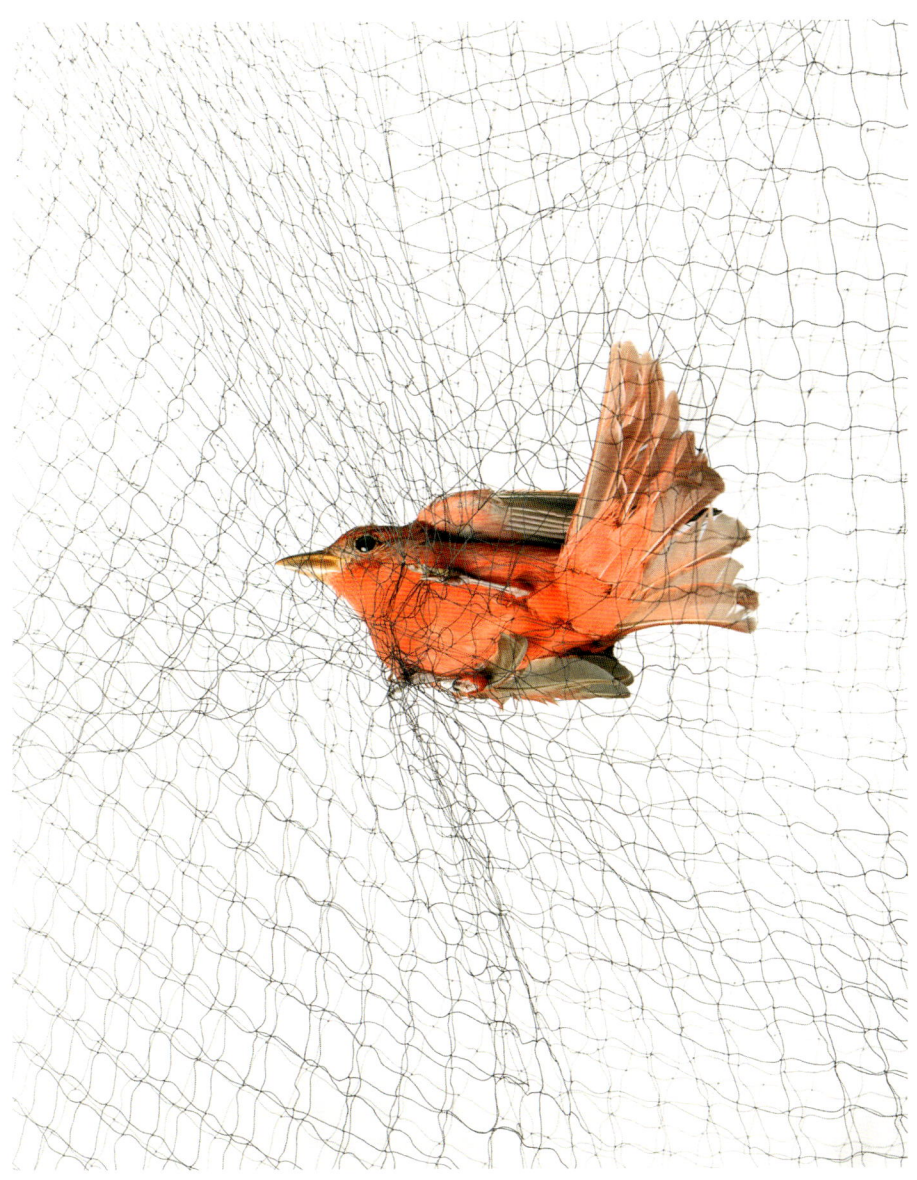

Summer Tanager
Piranga rubra

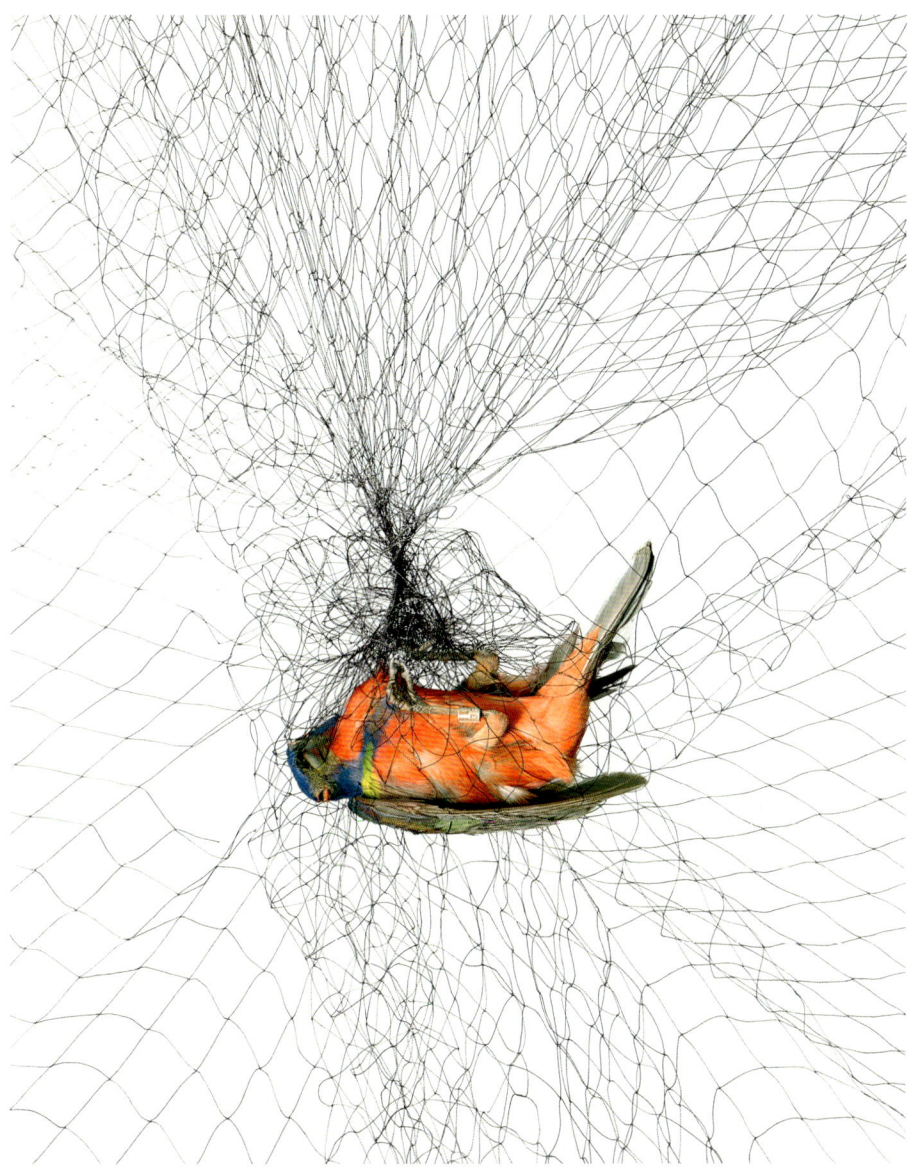

Painted Bunting
Passerina ciris

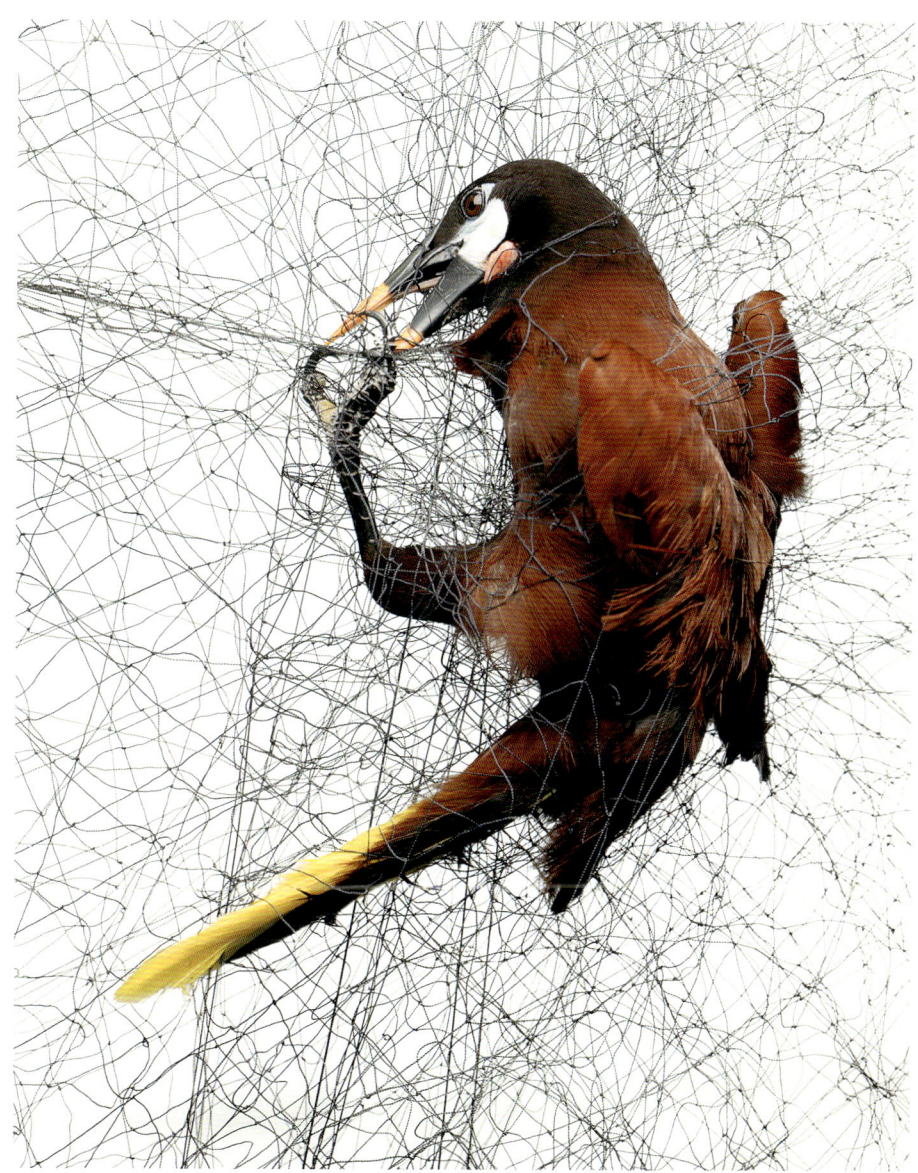

Montezuma Oropendola

Psarocolius montezuma

NOT STRICTLY FOR THE BIRDS
The Aesthetics and Ethics of Todd Forsgren's "Ornithological Photographs"

John A. Tyson

At first glance, Todd Forsgren's "Ornithological Photographs" seem a world away from Barnett Newman's relatively spare abstract expressionist paintings. In much of Newman's oeuvre, geometry coexists with the brushy traces of the artist's hand, and fields of color are split by vertical lines—dubbed "zips" by the artist—with which, in the gallery, spectators often align themselves. The painter reflected on this interplay between theory and practice, creative expression and science, at the Woodstock Art Conference in August 1952: "I feel that even if aesthetics is established as a science, it doesn't affect me as an artist. I've done quite a bit of work in ornithology; I have never met an ornithologist who ever thought that ornithology was for the birds." Newman rapidly condensed this statement into his classic quip: "Aesthetics is for the artist as ornithology is for the birds."[1] While this snappy line implied that artists were natural, creative creatures, it was also born out of a real interest in birds. During the early 1940s, Newman did not paint, but instead dedicated himself to the natural sciences: he attended classes at the American Museum of Natural History, took summer courses in botany and ornithology at Cornell University, joined the American Ornithologists' Union, and, accompanied by his wife Annalee, participated in an Audubon Society Camp in Newcastle, Maine.

It was in the same region of Maine that, some sixty years later, Todd R. Forsgren began to pursue both biology and art, though he had long been interested in the study of birds. Soon after graduating from Bowdoin College, Forsgren began work on his series "Ornithological Photographs" (2006–2014), which pictures manifold varieties of birds trapped in the grid lines of mist nets. Each photograph contains a single, temporarily caught specimen, fixed first in the net and subsequently by the artist's camera. In this series, Forsgren grapples with the legacy of John James Audubon—the namesake of the conservation society that organized Newman's excursion[2]—both paying homage to and deconstructing the problematic, highly composed depictions found in the classic *Audubon's Birds of America* (1827–1838). Yet Forsgren's series might also serve as a contemporary riposte to Newman, to whose words and works his project is a productive foil. The "Ornithological Photographs" series does not encourage the literal positioning of Newman's "zips," but Forsgren's crisp, detailed images demand that one take a

---

1 Newman, "Chronology," The Barnett Newman Foundation, accessed May 19, 2015, www.barnettnewman.org/chronology.php

2 The programming of the Bowdoin College Museum of Art perhaps contributed to Forsgren's awareness of Audubon's actual relation to nature: Walton Ford, who critiques the legacy of Audubon via very different means, had a solo show at the museum in 2000 titled *Brutal Beauty*.

stance, prompting viewers to contemplate ethics as well as aesthetics: "I'm trying to tangle the viewer as much as the birds are tangled," the artist affirms.[3] At the heart of this contemplation are the multicolored feathered creatures that occupy roughly the center of each photograph. Some, like the White-crowned Sparrow (*Zonotrichia leucophrys*), are quite common, while others, such as the Keel-billed Toucan (*Ramphastos sulfuratus*), seem, to North American eyes, exotic. Each is set against a white background crisscrossed by the fine warped grid of a mist net. In their representation and repetition of the supposedly rational grid, Forsgren's photographs evoke a history of geometric abstraction—rather like the paintings of the elder abstract expressionist, pushing against this formal structure even as they reproduce it.

Captivating and beautiful though they are, Forsgren's images in fact document a moment central to avian study. During the short period they occupy the nets, the birds' small, ensnared bodies are disciplined—alternately held in place or marked and measured by biologists—while the embedded photographer intervenes in the stream of scientific protocols with artistic operations such as the snowy backdrop. It is difficult to view the birds without considering how they came to be in the situation made visible, and even what they might be feeling at the time. Forsgren's "Ornithological Photographs" exist within and beyond established taxonomies—deriving art from science and catalyzing the contemplation of animal ontology within ornithology. Thus Forsgren's series asks us to reconsider epistemological categories, showing that the "shifting sands" of knowledge can never be captured in a single paradigm or bounded by one discipline alone.[4]

Field notes: Birds in the history of art and criticism

While aesthetics and ornithology may have preoccupied few modern artists beyond Newman—and even fewer birds—historians and theorists of art dating at least to Pliny the Elder have linked these creatures with the visual arts. The writer of *Naturalis Historia* recounts the ancient myth of the painter Zeuxis, whose *Child Carrying Grapes* depicts the fruit with sufficient naturalism that hungry birds begin to peck at the work's surface.[5] Audubon, along with Rev. John Bachman, contributed to an update of the Zeuxian trope, overseeing an experiment on the eyesight of vultures in which a painting of a carcass was found to attract the birds as effectively as real carrion.[6]

---

3 Todd Forsgren cited in Nick Lund, "The Birdist: Interview with Todd Forsgren, Photographer," *TheBirdist.com*, February 19, 2013, accessed November 30, 2014, www.thebirdist.com/2013/02/interview-with-todd-forsgren.html.

4 Forsgren, comment emailed to author, May 5, 2015.

5 For an account of Zeuxis in relation to beauty see Elizabeth Mansfield, *Too Beautiful to Picture: Zeuxis, Myth, and Mimesis* (Minneapolis: University of Minnesota Press, 2007), 26–27.

6 See John James Audubon, *Ornithological Biography, or An Account of the Habits of the Birds of the United States of America: Accompanied by Descriptions of the Objects Represented in the Work Entitled The Birds of America, and Interspersed with Delineations of American Scenery and Manners*, vol. II (Philadelphia, PA, and Edinburgh, Scotland: Judah Dobson; A. Black, 1831–1849), 44–46. A more mythic version of this story is presented on PBS's profile of Audubon. See Ken Chowder, *John James Audubon: Drawn from Nature*, PBS.org, July 25, 2007, accessed March 16, 2015, www.pbs.org/wnet/americanmasters/episodes/john-james-audubon-drawn-from-nature/106/

The influential nineteenth-century art critic John Ruskin was also a bird-watcher. Ruskin's interests spanned real birds; their representation in art, literature, and science; and the etymology of both specialized and vernacular words related to birds. Although he was a skeptical consumer of mainstream ornithology, Ruskin admired Audubon, whose prints—and a copy of *Ornithological Biography, or An Account of the Habits of the Birds of the United States of America*—he owned.[7] Beginning in 1873, he gave a series of talks at Oxford University on avian themes, each focusing on a single species of feathered creature.[8] Ruskin's lectures on birds are irreverent, taking paths that the critic was not willing to blaze in relation to works of art and their institutions, and in "Ornithological Photographs," Forsgren poetically traverses some of their rhetorical trajectories, while training a critical lens on the history and conventions of art and photography. Ruskin, for instance, identifies the sometimes irrational nature of scientific taxonomy—"The scientific classifiers are not to be beaten. If they cannot find a number of similar birds to give different names to they will give two names to the same one"[9]—and notes the problem of categorizing wildlife by "national school." He muses of the robin, "Where does he come from? I stated that my lectures were to be on English and Greek birds; but we are apt to fancy the robin all our own. How exclusively, do you suppose, he really belongs to us?"[10] Forsgren's series, similarly, is not limited to the birds of a single nation and—as will be discussed in due course—also indexes the tendency of "scientific classifiers" to name and rename. Ruskin employed stuffed specimens as props in his lectures; like Forsgren's photographs, these prompted reflections on the ethical and epistemological problems of collection and classification. Discussing a blue-breasted robin, Ruskin noted that "the last was seen shot at Margate…. This, then, is the utmost which the lords of land, and masters of science, do for us in their wrath upon our feathered suppliants. One kills them, the other writes classifying epitaphs."[11]

Audubon's legacies and the fine art of birding

Although today, the name Audubon is associated with the protection or conservation of nature, John James Audubon was to a great degree an ornithologist of the stripe described by Ruskin, most concerned with shooting specimens for collection. Indeed, paintings of Audubon himself, such as an 1826 portrait by John Syme, present him as a ruggedly handsome frontiersman: dressed in animal skins, rifle at the ready. Perhaps literalizing the French or Spanish term for "still

---

7 On Ruskin's skepticism of modern ornithology, see John Ruskin, *The Works of John Ruskin*, Vol. 29, E.T. Cook and Alexander Wedderburn, eds. (London and New York: George Allen; Longmans, Green, and Company, 1906), 509. On Audubon, see Ruskin, *Works*, Vol. 25, 181. In a rather curious coincidence, Oxford's Ruskin School of Art (founded in 1899) is housed in a building previously known as the Rookery.

8 These talks were subsequently published in the United States under the title *Love's Meinie: Lectures on Greek and English Birds* (New York: John Wiley & Son, 1873).

9 Ruskin, *Love's Meinie*, 9.

10 Ibid, 17.

11 Ibid, 13.

life" (*nature morte/naturaleza muerta*: "dead nature"), Audubon was a master fabricator of nature who honed his taxidermy skills while constructing displays for a museum of natural history that he briefly operated.[12] Yet his magnum opus was *Birds of America*. The ambitious project has fanciful dimensions, containing a number of images of birds with unnaturally contorted bodies (manipulated into these poses postmortem), and juxtaposing species that do not normally interact. Nevertheless, aspects of Audubon's work hinged on close observation of the natural world. *Birds of America*'s primary images are set against an austere, milky background, and the polychrome figures are often accompanied by diagrammatic components connoting scientific veracity, such as labeling and detail images of distinct avian body parts. Hence, while Audubon's birds fall between fact and fiction, the general public viewed them as truthful and accurate.[13]

According to Forsgren, *Audubon's Birds of America* was part of a "romantic quest, with the goal of painting (and shooting) all the birds in America," a goal paralleling that of the photographer (and many birders).[14] However, Forsgren acknowledges the impossibility of this objective, noting that would be extremely difficult for a bald eagle to settle in the nets.[15] As a result, his series might be viewed as a satire of Audubon's overblown aims, in which the example of the eagle (a metonym, of course, for the United States, spotted more frequently on seals and coins than in the wild) is conspicuous.[16] Indeed, Forsgren—whose subjects hail from the Americas, not solely the U.S.—does not imagine an avian body politic coextensive with national identity, and implicitly questions how Audubon reached his final count for *Birds of America*. For, as Ruskin points out, many birds are migratory—precisely how is their nationality determined?[17]

Forsgren's impossible-to-realize intention is close in spirit to conceptual artist Douglas Huebler's *Variable Piece #70* (1971–1997), the aim of which was to photograph all people presently alive. Huebler's awareness of his project's futility allowed him a certain degree of freedom in its realization. *Variable Piece #70* pokes fun at both utopian hopes that photographs could connect all of humanity (à la Edward Steichen's *Family of Man*) and dystopian visions of a totalizing archive containing everyone's data and image. Rather like Huebler's, Forsgren's photographic project evokes the human desire to collect an entire set of something—whether it

---

12 See William Souder, *Under a Wild Sky: John James Audubon and the Making of Birds of America* (Minneapolis: Milkweed Editions, 2014).

13 See Linda Dugan Partridge, "Audubon and the Tradition of Ornithological Illustration," *Huntington Library Quarterly*, Vol. 59, No. 2/3 (1996): 269–301.

14 Todd Forsgren, "Ornithological Photographs: Artist's Statement," accessed online March 17, 2015, www.toddforsgren.com/birds/index.html

15 Ibid.

16 Coincidentally, the number of birds in Audubon's volume coincides with the current number of representatives elected to the United States House of Representatives.

17 A nativist spirit of a distinct sort runs through the Audubon Society as well, which from early on advocated against the introduction of "foreign" species.

be birds seen or artworks purchased—and, by focusing on a uncontainable set, simultaneously stymies this totalizing impulse. The scope of Forsgren's series is not limited to birds of a nation, to specific species of birds, or to birds that occupy a common habitat. Hence the collection is simultaneously boundless and incomplete. The malleable, expanding structure of "Ornithological Photographs" mirrors that of the field of ornithology.

Huebler also engaged with birding in *Duration Piece #5* (1969). For this, he followed birdcalls in Manhattan's Central Park, taking a photograph as he walked in the direction of the chirps and whistles. Following these sounds was in fact the project's focus, for—deemphasizing both vision and the desire for scientific categorization—Huebler did not actually document any birds. Forsgren takes almost the opposite tack to make a similar point. The titles of the "Ornithological Photographs" are tautological and culled from biology: the common and Latin names of the birds, identifying their genus and species, are also used by Audubon and birders' field guides. However, in the case of more than a dozen of the titles, the initial Latin and common designations became outmoded; since the photo was taken, the bird is no longer to understood to fit into the animal kingdom in the same fashion. For example, the Clay-colored Robin is now the Clay-colored Thrush (*Turdus grayi*); the Magnolia Warbler (*Setophaga magnolia*), Chestnut-sided Warbler (*Setophaga pensylvanica*), and Adelaide's Warbler (*Setophaga adelaidae*), which all belonged to *Dendroica*—a genus that has been eliminated—are now all *Setophaga*. Nevertheless, on the artist's website or in older documentation of his exhibitions, many of the formerly correct names persist. As such, Forsgren's images might teach lessons that parallel the critical insights of Michel Foucault, calling for us to analyze the way that "words and things (next to and also opposite one another)…'hold together.' "[18] Continually augmented, knowledge is ever changing, its boundaries porous.

Again, rhyming with Huebler's *Duration Piece #5*, Forsgren probes the connections between seeing and knowing. Forsgren's series challenges birders to identify the specimens on display and rattle off their proper names—"Violet-crowned Hummingbird (*Amazilia violiceps*)"… "Blue-winged Warbler (*Vermivora cyanoptera*)"… "*Zeledon's* Antbird (*Myrmeciza zeledonia*)."… Such identifications are driven by faith in sharp eyesight, similar to the faith in vision (and taste) that drove many of the critics (Clement Greenberg, Michael Fried) who championed the mid-century abstract painting of artists such as Newman. The notion of the critic as possessed of an exceptional optical sensibility is also, moreover, associated with Ruskin, whose writings emphasize the capacity to see and discern in art.[19] Yet rather than reward viewers for good taste or familiarity with art history, Forsgren's series seems to privilege or at least appeal to the knowledge prized by the subculture of birders.

---

18 Michel Foucault, *The Order of Things* (New York: Random House, 1970), xviii.
19 For more on this idea, see Rosalind Krauss, *The Optical Unconscious* (Cambridge, MA: MIT Press, 1993).

The Audubon Society's mission is, in the organization's own words, "to conserve and restore natural ecosystems, focusing on birds, other wildlife, and their habitats for the benefit of humanity and the earth's biological diversity."[20] Conserving and restoring are also traditional functions of the art museum. While these goals are not necessarily negative, they are predicated on striving to fix their objects in an "authentic," ideal moment of conservation. Even the *Oxford English Dictionary*'s definition of "birding" ("the observation of birds in their natural habitat as a hobby") contains certain assumptions about authentic viewing in nature.[21] In these institutional missions and hobbies, moreover—building on the ideas of ecological theorist Tim Morton—nature, rather like art, is presented as autonomous of human existence, and in an elegiac form. "Traditionally," Morton writes, "elegies weep for that which has already passed. Ecological elegy weeps for that which will have passed given a continuation of the current state of affairs."[22] Seeming to confirm Morton's theory, there is presently a message on the Audubon Society's website that reads: "Nearly half of our birds are at risk of extinction this century."[23] Surely, statements like this one appeal to the hearts of readers, prompting donations. Forsgren, however, avoids this vision of nature, and does not photograph any endangered species.

Indeed, eschewing both the constructed nature of Audubon's images and the mourned, idealized environment of the Audubon Society, Forsgren's work clearly depicts a human-animal encounter. The series confronts spectators with the perhaps uncomfortable idea that we are part of the same ecological system as the birds he photographs, and with the surroundings whited out, the small, wild bodies are thrown into high relief.[24] In some sense resembling Audubon's drawings, Forsgren's birds' bodies are at times bent, twisted, and partially decontextualized. Nevertheless, contra his predecessor's seamless illusions, the photographer takes measures to reveal the artifice of the net, which draws out the wings and talons. Moreover, Forsgren provides no synecdoches of habitat. Gesturing to a parallel process of abstraction in his photographic operations and the work of the biologists he accompanies, the artist describes the instant he shoots as the "fragile and embarrassing moment before [the birds] disappear back into the woods, and into data."[25]

---

20 Audubon Society, "About Us," accessed March 16, 2015, www.audubon.org/about.

21 *Oxford Dictionaries Online*, s.v. "Birding," www.oxforddictionaries.com/us/definition/american_english/birding

22 Tim Morton, "The Dark Ecology of Elegy," in *The Oxford Handbook of the Elegy*, ed. Karen Weisman (Oxford University Press, 2010), 254.

23 Audubon Society, "About Us."

24 For more on this idea, see Morton, "Dark Ecology of Elegy," 252–253, and *Ecology without Nature* (Cambridge, MA: Harvard University Press, 2007), 1–28. This kind of interaction with birds in some ways recalls Hans Haacke's *Live Airborne System* (1966), which consisted of the artist feeding Wonder Bread to seagulls at Coney Island. The fact that the food and location are hardly the epitome of natural prompts reflections similar to those sparked by Forsgren's backdrop and nets. The only trace of this work, which Haacke considers a kind of performance, was a photograph.

25 Forsgren quoted in Justine Aw, "Todd Forsgren's Ornithological Photographs 08.09.12," *Notcot.com*, September 8, 2012, accessed March 15, 2015, www.notcot.com/archives/2012/08/todd-forsgrens-ornithological.php

The provocative ambiguity in Forsgren's statement raises questions about subject-object relations: is it the birds, or those who gaze upon their prostrate bodies, who feel embarrassed?

From the "art-science" to photographing the scientific process

Peter Henry Emerson's treatise *Naturalistic Photography for Students of the Art* (1890) is a manifesto in favor of photography's artistic status. Laying the groundwork for the theories of Alfred Stieglitz, Emerson envisions and emphasizes medium purity, arguing against manipulated negatives or the creation of composite images and advocating pure, artistic images. Despite his tome's title, the photographer acknowledges the fact that photography "is an art founded upon science."[26] Yet Emerson is definitive in censuring a term for the medium used in the second half of the nineteenth century: "the art-science." It is precisely this "improper" hybrid term that seems to encompass Forsgren's work, which—though it arguably hews closer to art—revives the term's critical hyphen, prompting reflection on the history of photography.

Birds were key subjects in early photographs of bodies in motion. Eadweard Muybridge, the nineteenth-century landscape photographer best known for his studies of human and animal locomotion, used a chronophotographic method to capture movement over time—most famously in his depictions of the racehorse Sallie Gardner galloping, at times with all four hooves off the ground. Muybridge also photographed birds, humans, and other animals. In the background of nearly all of his animal locomotion studies is a white grid on a black background that both enables the measuring of body part displacement and provides the images with a consistent aesthetic implying science and rationalism.[27]

Muybridge's experiments with freezing movement also inspired French scientist Étienne-Jules Marey. Marey was particularly concerned with birds in flight. Like Muybridge, he used chronophotography to capture their movements; Marey then created models of different phases in bronze and plaster. These sculptures were placed in a circle in a zoetrope—a drum with slits—which, when set in motion, provided viewers an illusion of a flying bird. Diagrams of this technique continue to circulate in textbooks on the history of photography, situating birds as one of the ideal subjects for scientific study via photography. Photographic and military technology nearly intersect in Marey's methods, which relied on gun-like devices. As the verb common to projectile weapons and cameras, "shoot," implies, the instruments share a position in the Western imagination. And though it does not cause the same bodily harm, shooting nature photographs might be driven by an impulse parallel to hunting: a desire to collect and fix subject-objects.[28]

---

26 P.H. Emerson, *Naturalistic Photography* (London: Sampson Low, Marston, Searle & Rivinton, 1890), 18.

27 See Rosalind Krauss, *The Originality of the Avant-Garde and Other Modernist Myths* (Cambridge, MA: MIT Press, 1986), 13.

28 For an extended discussion of these concepts see Paul Virilio, *War and Cinema* (London: Verso, 1989).

Forsgren's photographs capture birds frozen against gridded backgrounds, evoking the tradition outlined above. Moreover, while the nets in Forsgren's photographs can read as flattened, patterned ground, they are in fact dimensional, and almost equally legible as figures. This blurring of figure and ground recalls parallel operations in abstract expressionist paintings (as can be observed in Jackson Pollock's dripped skeins or Willem de Kooning's calligraphic all-overs). However, the photographer's series resonates most with 1960s-era minings of Muybridge by conceptual artists such as Sol LeWitt. Similarly, Forsgren's works suggest "scienticity" but do not really serve a scientific purpose: poetics, instead, is derived from biology. Framing the birds against white unites the images, linking the units in the series with common form and methodology.

Nevertheless, a tension between unity and differentiation emerges in the "Ornithological Photographs." The largely decontextualized birds recall the sitters in Richard Avedon's portraits: in both cases the photographic subject, set against a white backdrop, occupies the center of the frame. For the fashion photographer, this technique served to isolate and highlight the subject. Similarly, as Susan Wegner argues in this volume, we might consider Forsgren's images to be bird portraits. For although, as in *Audubon's Birds of America*, each feathered creature is a metonym for its species, the photographs' aesthetic operations push against the generic grain of the titles. Although the avian subjects do not pose, their behavior sometimes changes in the nets, highlighting their subjectivity: jays, for instance, birds that have a more "exacerbated character," become, as Forsgren notes, "docile in the nets."[29] The artist hopes that the series will prompt spectators to understand the diverse and multiple bodies not as merely signifying "birds," a tendency mitigated by the photographs' more intimate textual presentation.

Clearly occupying a space both in and between art and science, the "Ornithological Photographs" are not just artful renderings of nature. Forsgren's lens is trained not simply on birds but upon ornithology and the practice of science. His work is, on the one hand, the result of an aesthetic intervention in the process of biological research. Forsgren's minor interruption— adding a white background and taking a photograph—yields very different results from those of the teams of biologists who mark, weigh, sample, and measure, but his photographs are equally the result of fieldwork. Forsgren also sees parallels between his series and Sol LeWitt's wall drawings, both of which are driven by protocols, followed without deviation to their (il)logical conclusion.[30] A handful of LeWitt's mandates in his "Sentences on Conceptual Art" (1969) could almost read as "best practices" for data collection: "The process is mechanical and should not be tampered with. It should run its course."[31]

---

29 Todd Forsgren in conversation with the author, November 28, 2014.
30 Ibid.

31 Sol LeWitt, "Sentences on Conceptual Art," first published in *0-9* (New York, 1969), reproduced online, accessed March 17, 2015, www.altx.com/vizarts/conceptual.html

On the other hand, science and the scientific process are the subjects of Forsgren's work. Because the birds as photographic subjects closely resemble their real-world referents, the beholder is inserted into a scientific point of view and ethically implicated. Gazing on the shock of black and electric-yellow feathers in his photograph of a Boat-billed Flycatcher (*Megarynchus pitangua*), its ebony beak apparently open in a screech, yields visual pleasure punctuated with visceral discomfort. The process of turning bodies into data is not painless for all parties involved.

Grids to exceed rationality

Ernest Thompson Seton's duck diagrams, varieties of schematic waterfowl charted on a grid, were a great inspiration for the naturalist Roger Tory Peterson. *Peterson's Field Guide*, which was first published in 1934, helped to popularize birding; his name continues to be associated with all manner of guides. As this essay has already discussed, Muybridge relied on the perceived authority of grids to frame his photographic subjects. The strategy has enjoyed a rich series of lives and afterlives in the visual arts, particularly in modernist painting and in works that bridge media such as LeWitt's *Serial Project 1 (ABCD)* (1966). The grid is a central motif in many of Forsgren's photographic series, including the "Ornithological Photographs," which could be set into genealogies with all of the examples above.[32]

Following Rosalind Krauss, the grid is a structure that undergirds numerous avant-garde artworks (from those by Kazimir Malevich and Piet Mondrian in the first fifty years of the twentieth century, to works by Jasper Johns and Agnes Martin in the latter half). The grid resists perspective—the representation of recession—and does not map: "If it maps anything, it maps the surface of the painting itself."[33] Photographs, like high-modernist paintings, tend toward flatness. This rejection of depth through emphasis on the surface illuminates the tension between Forsgren's images and more typical shots of "nature," which are usually captured through unwieldy, "expensive, long, phallic lenses"—and hence imply a great depth of field.[34] The flat matrix in Forsgren's photographs seems to thwart telephotographic penetration.

Krauss furthermore affirms: "Flattened, geometrized, ordered, [the grid] is antinatural, antimimetic, antireal. It is what art looks like when it turns its back on nature."[35] In Forsgren's photographs, the grid is impacted by the unruly corporeality of natural beings. While the mist nets discipline the trapped bodies, the presence of the birds destabilizes the grid's own geometry. In contrast to the well-ordered lines and color fields of a Mondrian or Newman's

---

32 Both of Forsgren's series "Experimental Forests" (ongoing) and "Untitled Re:Iterations" (2000–2014) are structured around grids. Another series, "44% Blue" (ongoing), contains numerous images of gridded forms.

33 Krauss, *Optical Unconscious*, 10.

34 Todd Forsgren cited in Lund, "Interview," *TheBirdist.com*.

35 Krauss, *Optical Unconscious*, 9.

lines, zipping through colored ground, the living, polychrome, feathered balls derange the standardized, modular units. Krauss holds that the grid can be both centripetal and centrifugal, and indeed, in images such as Forsgren's photo of the Magnolia Warbler (*Dendroica magnolia*), his avian subjects warp the weave of the net such that its lines drive vectors of the gaze inward. Nonetheless, with its threads that extend to the edge of the frame, the net also unites the images, imagining them as woven out of the same web. The grid is often redoubled in the hanging of the "Ornithological Photographs" (or in its Internet presentation), suggesting that the series itself might expand ever outward. This configuration, in a sense, serves to ensnare the viewer, too, whose eye wanders along the thin black tendrils from bird to bird.

Grids, then, necessarily haunt the pages of this book. The volume you now hold, extending Forsgren's work, is a field guide for birding as conceptual art. The present expansion of "Ornithological Photographs" is particularly apt. For, like the series, it possesses the logic of the "para-": it is both within and beyond taxonomies. It is not strictly for the birds.

A BRIEF HISTORY OF BIRD-BANDING AND MIST-NETTING

James Lowen

It's early morning, and along a sixty-foot line dividing scrub from woodland, a mist net shimmers almost imperceptibly, eight feet high, stretched taut between poles. In it lies a feathered bundle, striped black and white. It is contorted but still, the nylon mesh tangled between its legs. This is when two ornithologists and photographer Todd Forsgren enter the scene. After deftly extracting the bird and slipping it into a breathable cotton bag, the trio returns to a nearby clearing, where they have established a temporary research station. Within minutes, the Black-and-white Warbler is measured and weighed, and its leg is encased in a tiny aluminum anklet. The bird will return to the wild to resume its life, while the data the scientists have gathered will go on to shed light on bird survival and movement, and on our changing environment.

The history and science of banding

The concept of banding originated with John James Audubon. In 1803, Audubon conducted the first bird-marking experiment by attaching silver thread to the legs of Eastern Phoebes.[1] Two of the flycatchers returned to the same site the following spring, offering the first-ever insights into birds' survival and site-fidelity. Within a century, banding schemes were operational across North America and Europe.[2,3] Uniquely numbered metal bands turned anonymous birds into recognizable individuals, available to help unravel avian mysteries relating to population trends, classification, and migration.

The invention of mist nets in the 1950s provided ornithologists with an efficient, safe way to capture birds for banding, and there are now more than 6,000 qualified banders in North America alone, with a coherent work program extending from the Canadian Arctic to the Latin American tropics.[4] Over the program's lifetime, four million marked individuals have been recovered or reported.[5] The resulting large-scale, long-term data set offers remarkable insights into birds' movements, populations, and survival.

1 See Rebecca Olsen, *Audubon's Aviary: The Original Watercolors for the Birds of America* (New York: Skira Rizzoli, 2012).

2 British Trust for Ornithology, "History of Ringing," 2015, accessed March 23, 2015, www.bto.org/volunteer-surveys/ringing/about/history-ringing.

3 John Tautin. "One Hundred Years of Bird-Banding in North America," USDA Forest Service General Technical Report, PSW-GTR-191 (2005).

4 Tautin, "One Hundred Years of Bird-banding."

5 United States Geological Survey, "How Many Birds Are Banded?" (2011), accessed March 23, 2015, www.pwrc.usgs.gov/bbl/homepage/howmany.cfm

Our understanding of migratory routes, stopover sites, and wintering areas largely originates in banding. In 1912, for example, a recovery from South Africa debunked the centuries-old myth that Barn Swallows wintered under water.[6] Chimney Swift wintering grounds were unknown until 1944, when bands from shot birds were handed in at the United States Embassy in Peru.[7] Shorebirds such as the Red Knot, marked with colored bands, can be tracked by bird-watchers along their migratory flyways, enabling conservationists to pinpoint key refueling stations in need of protection.[8]

Among the species photographed by Forsgren, banding recoveries have provided numerous revelations. White-crowned Sparrows were found to migrate along one of three routes (coastal, intermountain, or prairie) and to travel up to 183 km per day.[9] A Swainson's Thrush banded in Pennsylvania in 1966 was killed by a local hunter's blowgun near Sinchi-Yacu, Peru, some 3,150 miles away.[10] A banded Black-and-white Warbler trapped in Maryland in 2013 was found to be at least eleven years old, a ripe old age for such a small creature.[11]

Although we have much to learn about migration, today's banding schemes have shifted focus. Ornithologists now seek to examine the effects of environmental issues such as climate change, pollution, and habitat conversion on bird populations. Accordingly, banders' focus has switched to monitoring survival—the proportion of birds that endure the stresses of breeding, migration, and severe weather to persist between years.[12] The data are troubling: the North American Monitoring Avian Productivity and Survivorship (MAPS) program suggests that bird populations overall are declining, on average, by 1.77% per year.[13] An unfettered extrapolation of that rate would mean there would be no birds left at all by 2075.

---

6 British Trust for Ornithology, "History of Ringing."

7 Department of the Interior, untitled press release, 1944, accessed March 23, 2015, www.pwrc.usgs.gov/bbl/homepage/FredLincolnpressrelease.pdf

8 See Brad A. Andres, "Contributions of Bird Banding to International Waterbird Conservation," in *Birding Banding in North America: The First Hundred Years*, eds. Jerome A. Jackson, William E. Davis Jr., and John Tautin. (Cambridge, MA: Memoirs of the Nuttall Ornithological Club 15, 2008).

9 Barbara B. DeWolfe, George C. West, and Leonard J. Peyton, "The Spring Migration of Gambel's Sparrows through Southern Yukon Territory," *Condor* 75(1973): 43–59.

10 Robert C. Leberman and Mary H. Clench, "Bird-banding at Powdermill, 1971." Research report no. 30. Pittsburgh: Carnegie Museum of Natural History (1972).

11 Manomet Center for Conservation Sciences. "Spring Banding Season Underway, 11 year-old Manomet Bird Recaptured in Maryland," 2013, accessed April 28, 2015, www.manomet.org/newsletter/spring-banding-season-underway-11-year-old-manomet-bird-recaptured-maryland

12 See Jerome A. Jackson, William E. Davis Jr., and John Tautin, eds. *Birding Banding in North America: The First Hundred Years* (Cambridge, MA: Memoirs of the Nuttall Ornithological Club 15, 2008).

13 David F. DeSante and Danielle R. Kaschube, "The Monitoring Avian Productivity and Survivorship (MAPS) Program, 2004, 2005, and 2006 Report," *Bird Populations* 9(2009): 86–169.

Banding has alerted conservationists to the need for international cooperation. Wood Thrush populations, for example, seem to be declining because the birds, for unknown reasons, succumb on Central American wintering grounds. Only cross-border collaboration can possibly assure them a safe future.[14] The same may be true of many other species photographed by Forsgren, such as Blue-winged, Chestnut-sided, and Magnolia Warblers. At the same time, banding data have already helped conservationists restore populations of endangered species such as the Bald Eagle, Whooping Crane, and Californian Condor.[15] The uses of information from banding projects extend beyond conservation as well: the Adaptive Hunting Management Program, with banding data as a key input, enables the calculation of annual hunting quotas for waterfowl.[16] Tracking daily movements of banded Red-winged Blackbirds has helped identify management efforts to reduce crop depredation.[17] Banders follow the avian carriers of encephalitis, malaria, and West Nile virus; for the latter, research can now predict the timing and scope of future epidemics.[18]

Ethics

Such benefits could not have been gained without uniquely marking birds. But do such ends really justify the means? Or does trapping a bird in a net constitute unacceptable interference with a wild creature? Evidence points to the procedure's safety, with an independent survey revealing that six birds in every thousand are injured during mist-netting, and two per thousand die.[19] Given capture rates, this means that perhaps 2,000 birds perish in North America each year as a result of banding, a tiny fraction compared to an estimated 39 million birds killed by cats annually—in Wisconsin alone.[20]

Banding, likewise, does not appear to disturb bird routines: approved handling techniques minimize stress, enabling birds to go back to normal life shortly after release.[21] Studies confirm that breeders rapidly return to eggs or chicks, and that long-distance migrants continue their

14 James F. Saracco, J. Andy Royle, David F. DeSante, and Beth Gardner, "Modeling Spatial Variation in Avian Survival and Residency Probabilities," *Ecology* 91(2010): 1885–1891.

15 Jackson, Davis, and Tautin, eds. *Birding Banding in North America*.

16 John Tautin, Lucie Metras, and Graham Smith, "Large-scale Studies of Marked Birds in North America," Bird Study 46(1999): 271–278.

17 Richard A.Dolbeer, "Movement and Migration Patterns of Red-winged Blackbirds: A Continental Overview," USDA National Wildlife Research Center Staff Publications, Paper 152 (1978).

18 Jackson, Davis, and Tautin, eds. *Birding Banding in North America*.

19 Erica N. Spotswood, Kari Roesch Goodman, Jay Carlisle, Renée L. Cormier, Diana L. Humple, Josée Rousseau, Susan L. Guers, and Gina G. Barton, "How Safe Is Mist Netting? Evaluating the Risk of Injury and Mortality to Birds," *Methods in Ecology and Evolution* 3(2012): 29–38.

20 See John S. Coleman, Stanley A. Temple, and Scott R. Crave. *Cats and Wildlife: a Conservation Dilemma* (Madison: University of Wisconsin, 1997).

21 See *The North American Banders' Study Guide*, North American Banding Council Publications (2001).

journeys.[22] When properly secured, the weight ratio of the anklet itself—lightweight and spinning freely around the leg—is similar to that of a wristwatch on a human.

Such low incident levels seem to be the welcome result of stringent controls on those who wish to band birds. Catching birds without a permit is illegal in both North America and Europe, and securing a license requires intensive training, typically over several years.[23] The welfare of the bird is paramount. The research and conservation organizations that act as supervisory bodies have a vested interest in such vigorous regulatory scrutiny, for by controlling who bands where and when, they position themselves at the epicenter of avian science. They provide, manage, and interrogate the repository of banding records. They design banding programs to address crucial conservation questions.[24] The organizations also guide banders in producing high-quality data with minimum adverse impact on the feathered subjects of study.

Every bird banded—and the information it subsequently divulges—is part of a bigger picture. That Black-and-white Warbler is, undeniably and joyously, a natural wonder in itself. Thanks to its tiny metal anklet, however, it will have much wider environmental resonance. Forsgren's portraits capture this duality—inviting empathy with the individual bird in all its splendor, while illustrating a deeper beauty: that of the value of scientific study, both for conservation and for our enhanced appreciation and understanding of the natural world.

---

22 British Trust for Ornithology, "A Bird in the Hand: Ringing Birds for Conservation," 2014, accessed March 23, 2015, www.bto.org/sites/default/files/u17/downloads/about/resources/birdtableinsert_highres.pdf

23 *North American Banders' Study Guide.*

24 Robert A. Robinson, Catriona A. Morrison, and Stephen R. Baillie, "Integrating Demographic Data: Towards a Framework for Monitoring Wildlife Populations at Large Spatial Scales," *Methods in Ecology and Evolution* 5(2014): 1361–1372.

SNARING THE VIEWER

Susan Wegner

Todd Forsgren's title for his series "Ornithological Photographs" has an understated and scientific cast that gives no hint of the pathos, the exquisite colors and forms, the intimacy and immediacy of his images of single birds caught in mist nets. Suspended in midair, snared in a mesh fine as a spider's web, caught in angry alarm or in a serene dancer's pose, the many variations of Forsgren's portraits spark an equally wide range of associations in the viewer's memory.

And they really are portraits. They present unique, specific members of a species, not some ideal or composite. These images catch the birds in imperfect poses that are dependent on their weight, the speed with which they encounter the net, and the force of struggle they put up trying to free themselves. They are portraits after impact, with all the accidental results. Sometimes the net strike yields a pure side view of a bird, giving the distinctive profile favored by field guides. More often, chance and physics create poignant distortions, as in the case of the Painted Bunting hanging belly up, held fast by one leg tangled in a dense cone of threads. With its head foreshortened to a blue nub, one wing cupping its body like a turtle's carapace, the bunting loses most of the visual cues of its "birdness."

Forsgren's mist-net portraits of particular beings stunned by their own minute disasters might even bear comparison with Dorothea Lange's portraits of men and women trapped in dust-bowl Texas. Lange's photographs, such as those of Nettie Featherston, vividly present a lone figure confronting an incomprehensible calamity. Featherston's form is reduced to essence, arms bent sharply at the elbow as she clutches her brow and shoulder convulsively. Forsgren's portraits of a bird's instinctive struggle carry similar power. Created in a sliver of a second, these works repay close and unhurried looking.

Still-life painting and early bird photography

Forsgren's works take part in a long history of bird images within European and American art and science. For centuries, wealthy patrons could enjoy one of the major subsets of still-life painting, works featuring the bodies of recently slain game birds. With ducks and grouse hanging head down, smaller prey strewn below, these works celebrated the unseen hunter's prowess and anticipated the delights of the table to come. The "game piece" reached heights of verisimilitude in the seventeenth-century Lowlands, with specialists such as the Flemish painter Jan Fyt (1611–1661), one of whose characteristic oil paintings, "Still Life with Dead Birds," hangs in the Museum of Art at Bowdoin College, Forsgren's alma mater. With painstaking observation, the painter skillfully renders the varied textures of his feathered subjects, from the smooth, close-set glossy head feathers to the stiffly projecting flight feathers fanned out by the pull of gravity, to the soft puffs of down at throat and breast. The beak of the green-headed mallard aligns perfectly with the bodies of fallen snipes.

Closer in time, medium, and approach to Forsgren's efforts, Eadweard Muybridge's pioneering photographic examinations of animal locomotion document pigeons, cockatoos, and raptors in flight. Muybridge lined up a series of cameras along the birds' flight paths, effectively taking a sequence of stop-action images. A pigeon in flight, viewed from the side against a gridded backdrop, reveals the extensions and flexions in a single wing-beat. Even more revelatory, the same series shot head-on yields body silhouettes that range from a slim vertical through a butterfly spread to a stage where the wings are held so tightly along the line of the body that they almost disappear. Forsgren's "Ornithological Photographs" match the Fyt painting's delight in color, texture, and shape, and also share with Muybridge's work the unexpected, sometimes shocking distortions of a bird's body isolated against a squared grid and white backdrop.

Audubon's Birds of America

Forsgren has often stated that his twenty-first-century series continues and expands the work of North America's most famous bird artist, John James Audubon (1785–1851), whose ambitious double elephant folio *Audubon's Birds of America* (1827–1839) sought to picture at life size every bird then known within the region. The resulting 435 hand-colored etchings with engraving and aquatint still rank as a landmark in ornithological art.

Setting Forsgren's achievements against Audubon's is not meant to be a contest of greatness, but rather a way to highlight the remarkable originality and subtlety of the "Ornithological Photographs." Curators and critics writing on Forsgren's works starkly contrast the method of mist-netting to Audubon's ways of acquiring his subjects. Nineteenth-century ornithologists routinely shot their own specimens or collected ones already dead. Yet Audubon also spent years in the wild observing living birds, and he drew some captives that he had only slightly wounded in his hunting forays. Even his slain specimens were drawn quickly, soon after their death. The modern mist nets that furnish Forsgren his subjects are far less lethal than a blast of birdshot, although the nets also snag "by-catch," birds not the target of specific study. Forsgren photographs both these "accidental" birds and the scientists' subjects just before they are carefully extricated from the mesh, after which they are identified, measured, weighed, and quickly released. Forsgren's images suspend these flyers in the net that momentarily impedes them, holding them for our viewing, ultimately to be released alive.

Only an artist trained as a biologist could have put these birds before our eyes in this way. Forsgren understands the scientists' methods and tools, earning their trust to work his art within the structure of their study. With comparable patience, curiosity, and care, scientist and artist work in concert.

Implied human presence

Audubon's bird paintings and the hand-colored prints made after them for publication almost never reveal the mark of human activity. The birds reside in a fiction of their own private worlds, untroubled by a hunter's presence. At times, Audubon's collaborators, who produced settings for the finished prints, included a minor touch of the manmade world. For example, his principal printmaker, Robert Havell Jr., added an empty keg bobbing along the surface of the sea beneath Audubon's acrobatic Wilson's Petrels (*Birds of America*, plate 270). The petrel on the right flaunts its rump, ostentatiously displaying its undertail coverts to the viewer, and easily overshadows the background flotsam. On another sheet, a hazy row of distant structures and a fence line barely register beneath the dramatic snaky neck and curved bill of the Glossy Ibis (*Birds of America*, plate 387) on the hunt for prey.

Where Audubon erased or muted the human presence in his works, Forsgren insists upon our involvement. Here we are, inextricably tied to these captures: raw nature in collision with our civilization. Just as the unavoidable clash between seven billion of us and the winged world now presses many bird populations into decline, we create the conditions for these photographs: our machines manufacture the nylon nets; our hands string them. Forsgren's own ingenious mobile, out-of-doors studio—comprised of a large-format camera, a strobe with soft box, a bounce, and a white background—provides the pure neutral ground against which net and bird stand out. Its artificial lights banish time of day, atmosphere, and environment. The studio even suppresses the shadows of entangled bodies and twisted mesh that we expect to see cast onto the background. The results are carefully crafted, orchestrated artifices that pulse with the life of the candid moment.

These manmade devices veil the meaning of a bird's travels: a search for food or a mate, a long-distance migration, a patrol of territory, escape from a predator. What we lose of context we gain in undistracted focus. We see living beings, descended from dinosaurs, diminishing in number but still abundant. Even in cities, birds live around us yet go mostly unseen, as our vision is ever more riveted on ubiquitous little glowing screens. Forsgren redirects our gaze to miniature dramas of birds' lives, privileging patches of vivid color and irregular shape.

And we are there, close enough to look into the eye of the ensnared creature, close enough to read another mark of human interaction with the birds. A silver band gleams above the Black-and-white Warbler's toes, outdone by the brilliance of the liquid eye. The warbler's black, gray, and white markings add a matte counterpoint to these shining accents. In another image, an Adelaide's Warbler sports tiny black and silver tubular bands on its right leg. These warblers have been caught before, numbered metal crimped around their toothpick-thin limbs. The sized bands are calibrated to the birds' scale; the tiny warblers wear size zero.

Reactions to the net

Birds display a variety of behaviors when caught in a mist net. Species that are hard-wired to freeze in place when threatened hang passively, like insects bound in a spider's web. Those whose instincts urge them to fight or flee writhe in the nets, fluttering futilely. They call out in alarm. Some of Forsgren's portraits convey those struggles. The Common Ground-dove hangs like an inert puppet. One of its splayed tail feathers has been split by the net. A tight strand at the shoulder entwines its half-raised wings. It could almost be a twin to Muybridge's pigeon in flight in its most compressed silhouette, the clear grid distorted by the dove's efforts to escape.

The Variable Seedeater looks at us directly, regarding, assessing, recalling Audubon's few face-first portrayals. In his showy flock of now extinct Carolina Parrots (Carolina Parakeet), the lowest bird, head fully frontal, peers out in curiosity, aware. In another image a covey of Virginian Partridge (Northern Bobwhite) scatters underneath the outstretched talons of a Red-shouldered Hawk, with one bird lunging straight outward in terror. The Variable Seedeater seems almost philosophical in comparison.

The larger species display more agitation in the snare. The Boat-billed Flycatcher—at nine inches in length, one of the largest of the tyrant flycatchers—shrieks as it struggles. One strand of the net broken in the fight stands out against its yellow breast. A Rufous-winged Woodpecker hunches disconsolately, a wingless lump, feet effaced by a dense clump of tangled net. Distortions in the net resembling scribbled arcs suggest phantom wings, spreading above the captive.

Moved to learn more

Whereas birders' field guides strive for a typical static, profile image, Forsgren's images provide portraits of idiosyncratic individuals rather than perfectly defined "types" of the species. However, in one regard Forsgren's photographs do align with the field guides he admires, such as Robert Tory Peterson's revolutionary series, begun in 1934, that uses drawings to highlight field marks—a bird's distinctive spots of color, stripes, or patterns—for quick identification. Like the drawings in field guides, Forsgren's photographs are accompanied by the common English name and the Latin scientific name of the species. This link encourages us to go beyond the initial emotional punch and aesthetic joy of looking, and to venture toward gaining a deeper understanding.

Here are three examples of photographs and their titles that moved me to investigate further: two natives of the tropics and a familiar North American bird.

Puerto Rican Tody (*Todus mexicanus*), 2009

This tody is unique to the island of Puerto Rico, despite its scientific name, which is a mistake maintained from nineteenth-century labeling. Puerto Rican pride in this endemic avian gem bristles at the misnomer. This appealing little stub of a bird is affectionately named San Pedrito ("Little Saint Peter") as well as *mediopeso* (half peso) and *papagayo* (parrot). With a brilliant enamel-like green body, blood-red throat and lower beak, and yellow plumage aft, she—and she is a she, as signaled by the bull's-eye of her white eye ring—leans strongly forward in the net as if still in darting pursuit of her insect prey. The force of her forward stab into the web creates a wide vee of fabric behind her, like the shock waves that follow a jet aircraft.

Keel-billed Toucan (*Ramphastos sulfuratos*), 2013

This toucan regards us with soulful eye, its ornate beak folded close to its breast. The pose actually conforms with this gregarious bird's resting habits. Several toucans will mass together in tree cavities, where the birds will tuck their beaks and tails under their bodies so as to fit together as they sleep. Forsgren's photograph gives us a good view of the toucan's pale blue feet, with two toes forward and two backward (zygodactyl) furnishing a superior grasp on a branch. This, the charismatic national bird of Belize, is still hunted for its meat.

White-crowned Sparrow (*Zonotrichia leucophrys*), 2006

We conclude where Forsgren began, with a modestly colored bird of North American backyards, the White-crowned Sparrow, peering between upended feet in a pose insulting to a creature of aerial grace. The sparrow's plunge into the net matches the plunge of its numbers: though the species still boasts a large population and is thus marked as of least concern on Cornell Ornithology Laboratory's scale of endangerment, mist-netting has revealed a 33 percent decline in numbers over the last half century. Even common birds are under pressure.

Forsgren's photographs offer a potent antidote to our self-absorption and distraction, and invite us to witness, ponder, and care. His ornithological portraits bring us eye to eye with a fellow creature, putting a living face on the vast abstraction of "the environment." If they inspire us to learn about his avian subjects, then his images amply fulfill art's ancient charge to instruct as well as to delight.

References

Audubon, John James. *Birds of America* (London: Robert Havell Jr., 1827–1838)

Jan Fyt (attributed to), *Still Life with Dead Birds*, oil on canvas, 26 in. x 19½ in. Bequest of the Honorable James Bowdoin III, 1813.33. Bowdoin College Museum of Art, Brunswick, Maine.

Muybridge, Eadweard. "Pigeon flying," plate 756, *Animal Locomotion: An Electro-Photographic Investigation of Consecutive Phases of Animal Movements*. 1872–1885, published under the auspices of the University of Pennsylvania. Plates printed by the Photo-Gravure Company, Philadelphia, 1887.

Spotswood, E.N., and K.R. Goodman, J. Carlisle, R.L. Cormier, D.L. Humple, J. Rousseau, S.L. Guers, and G.G. Barton (2012), "How Safe Is Mist Netting? Evaluating the Risk of Injury and Mortality to Birds," *Methods in Ecology and Evolution*, 3: 29–38. doi: 10.1111/j.2041-210X.2011.00123.x

Cornell Ornithology Laboratory North American Breeding Bird survey, 1966–2010.

ACKNOWLEDGMENTS

Completing a photographic project and making a book can be quite a long process, and this certainly was the case with my *Ornithological Photographs*. Thankfully I had a lot of help during the ten years that it took me to produce this portfolio. I'd like to mention the people who helped me make the idea into a reality.

First, I would like to thank the tireless ornithologists who allowed me to accompany them in the field. Without their work, these photographs would never have been possible. These dedicated scientists include Grasiela Casas, Bento Collares, Ilma Dancourt, Timothy Guida, Landon Jones, Edye Kornegay, Elissa Landre, Alejandra Martinez, Fábio Melo, Giuliano Müller Brusco, Alejandra Pizarro, Márcio Repenning, Rafael Rueda, Angelina Ruiz Sánchez, and Julie West. More broadly, these photographs are meant as a celebration of the research of countless ornithologists; without their valuable work we would know far less about our planet's birds and their populations. Thank you as well to their subjects and my subjects: the birds who endured extra time in the net to be part of this project.

A number of friends and assistants lent an essential hand with the guerrilla studio photography needed to make these images. Thank you to Jenny Borth, Sam DiDonato, Lawrence Getubig, Daniel Goldstein, Anna Han, Ryan Johnson, Erini Koulouris, Matthew Moore, Conor O'Brien, Marc Redford, Angel Rueda, and Maureen Thompson. Their help in lugging and holding equipment was crucial. And the pleasure of their company as travel companions was an honor and a privilege.

Thank you, Michael Itkoff and Taj Forer at Daylight Books, for believing in the project enough to publish this volume, as well as Ursula Damm, Elizabeth Bell, and the entire Daylight team. Thanks to Jen Bekman of Jen Bekman Gallery, Joseph Carroll at Carroll and Sons, and Margaret Heiner of Heiner Contemporary. Without these gallerists' early support, I would have had a much harder time making and promoting these photos. Thanks to my writers, James Lowen, John Alistar Tyson, and Susan Wegner, for their thoughtful interpretations of the work. Thank you, Julian Montague, for the illustrations. Thanks to Alice Lovejoy for her effort shaping the texts for this book. To John McKee, thanks for convincing me to take photography somewhat seriously.

Finally, I'd like to thank my family. Brian and Suzanne Forsgren, my parents, encouraged me to develop an appreciation, compassion, and love for the natural world from an early age. My uncle, John Forsgren, has been a steadfast supporter of my work and life. My wife, Mika Yoshitake, is my biggest inspiration and my fiercest critic. She has provided me with incredible emotional support and holds me intellectually accountable.

Todd R. Forsgren
May 2015.

BIOGRAPHIES

Todd R. Forsgren uses photography to examine themes of ecology, environmentalism, and perceptions of landscape while striving to strike a balance between art history and natural history. Forsgren is currently a visiting professor at St. Mary's College of Maryland. He studied biology at Bowdoin College and photography at the School of the Museum of Fine Arts Boston and J.E. Purkyně University. He was an artist in residence at the Sitka Center for Art and Ecology and a Fulbright Fellow in Mongolia.

Brian W. Forsgren is the founder of the Gateway Animal Clinic, a veterinary practice in inner-city Cleveland that focuses on supporting the human-animal bond in the face of financial hardship. He was president of the Ohio Veterinary Medical Association's (OVMA) in 1999 and served as chairman of the Animal Welfare committee from 1996 to 1999. Dr. Forsgren has also been honored as Ohio's Veterinarian of the Year (2000) and received the OVMA Distinguished Service Award (1996). Named Distinguished Alumnus by the Ohio State University's College of Veterinary Medicine (2009), he also received the American Veterinary Medical Association's Leo K. Bustad Companion Animal Veterinary of the Year Award (2009) and the Humane Society Veterinary Medical Association's Veterinary Advocate of the Year Award (2011).

James Lowen is a wildlife and travel writer whose work opens up the knowledge of specialists in biology and bird-watching to a much broader readership. His books include visitor guides to the wildlife of Antarctica, Brazil, and Britain. In Lowen's previous career as a tropical conservation biologist, he mist-netted birds in Costa Rica, Indonesia, Madagascar, and Paraguay.

John A. Tyson is a specialist in modern and contemporary art. His scholarship primarily focuses on texts as platforms for exhibition, the relationship between conceptual art and performance, and the intersections of art and technology. Tyson recently defended his dissertation, "Hans Haacke: Beyond Systems Aesthetics," a project supported in 2014–2015 by a Henry Luce/ACLS Dissertation Fellowship in American Art. Tyson is presently the recipient of an Andrew W. Mellon Postdoctoral Curatorial Fellowship at the National Gallery of Art.

Susan Wegner teaches art history at Bowdoin College in Brunswick, Maine, specializing in the Italian Renaissance and seventeenth-century art. Recent projects focus on art and science, including the Bowdoin College Library exhibition Envisioning Extinctions: Art as Conscience and Memory (2014) and an AOU (American Ornithologists' Union) conference paper, "The Passenger Pigeon's Impact on North American Art and Culture: 1500–2014" (2014), both marking the centenary of the extinction of the passenger pigeon. Current research interests include assessing European bird species as models for angels' wings, and Audubon's rivalry with great ancient and European artists.

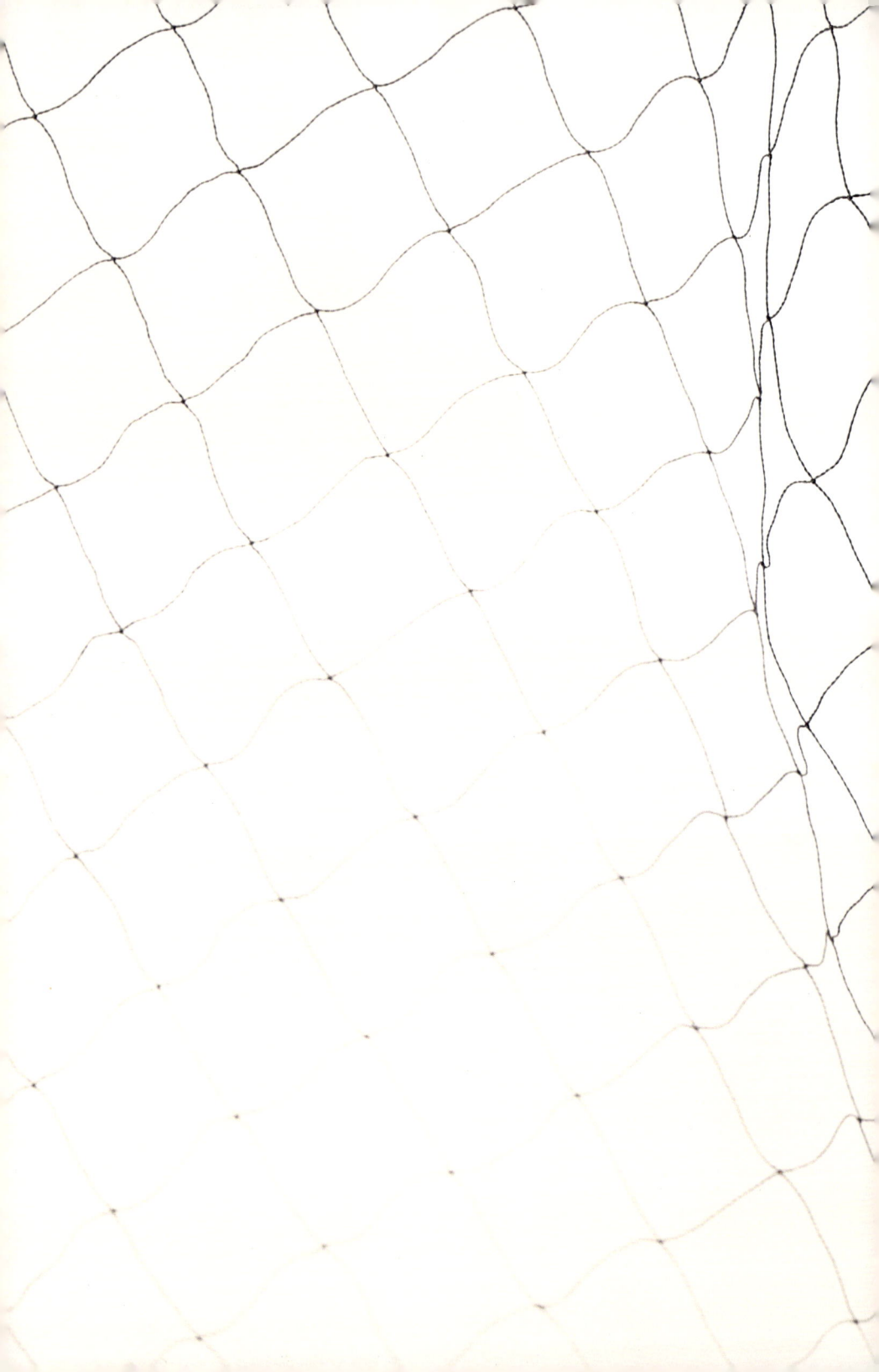